Energy Storage Devices - A Comprehensive Overview

Edited by Almoataz Y. Abdelaziz,
Mahmoud A. Mossa and Mohit Bajaj

Published in London, United Kingdom

Energy Storage Devices - A Comprehensive Overview
http://dx.doi.org/10.5772/intechopen.1003436
Edited by Almoataz Y. Abdelaziz, Mahmoud A. Mossa and Mohit Bajaj

Contributors
Ahmed G. Abokhalil, Amritha Kadathamana, Ananya Pal, Andrea Cerdán-Pasarán, Arya Abdolahi, Atanu Roy, Ciprian Cristea, Debasree Ghosh, Florica Şerban, Gopi R.R., J.A. Hernández-Magallanes, Jithin Varghese, Khairy Sayed, Kimia Shirini, Mahmoud A. Mossa, Mahmoud Aref, Maria Cristea, Pallavi Kumari, Radu-Adrian Tîrnovan, Sanal K.C., Shadai Lugo Loredo, Sina Samadi Gharehveran, Soorya Pushpan

First published in London, United Kingdom, 2025 by IntechOpen
IntechOpen is the global imprint of INTECHOPEN LIMITED, registered in England and Wales, registration number: 11086078, 167-169 Great Portland Street, London, W1W 5PF, United Kingdom

For EU product safety concerns: IN TECH d.o.o., Prolaz Marije Krucifikse Kozulić 3, 51000 Rijeka, Croatia, info@intechopen.com or visit our website at intechopen.com.

British Library Cataloguing-in-Publication Data
A catalogue record for this book is available from the British Library

Energy Storage Devices - A Comprehensive Overview
Edited by Almoataz Y. Abdelaziz, Mahmoud A. Mossa and Mohit Bajaj
p. cm.
Print ISBN 978-0-85014-721-6
Online ISBN 978-0-85014-720-9
eBook (PDF) ISBN 978-0-85014-722-3

If disposing of this product, please recycle the paper responsibly.

Meet the editors

Almoataz Y. Abdelaziz received the B. Sc. and M. Sc. degrees in electrical engineering from Ain Shams University, Cairo, Egypt, in 1985 and 1990, respectively, and the Ph.D. degree in electrical engineering according to the channel system between Ain Shams University, Egypt, and Brunel University, U. K., in 1996. He has been a professor of electrical power engineering at Ain Shams University since 2007. He has authored or co-authored more than 525 refereed journal and conference papers, 45 book chapters and 6 edited books with Elsevier, Springer and CRC Press in his research areas, which include the applications of artificial intelligence, evolutionary and heuristic optimization techniques to power systems operation, planning, and control. Prof. Abdelaziz is the chair of the IEEE (Institute of Electrical and Electronics Engineers) Education Society chapter in Egypt, editor of Electric Power Components & Systems Journal, Editorial Board member, Editor, Associate Editor, and Editorial Advisory Board member for many international journals. He is also a senior member of IEEE, a member of IET(Institution of Engineering and Technology) and the Egyptian Sub-Committees of IEC (International Electrotechnical Commission) and CIGRE (International Council on Large Electric Systems). He has been awarded many prizes for distinct research and international publishing from Ain Shams University and Future University in Egypt.

Mahmoud A. Mossa received bachelor's and master's degrees in electrical engineering from the Faculty of Engineering at Minia University in Egypt in 2008 and 2013, respectively, and the Ph.D. degree in electrical engineering in April 2018. Since January 2010, he has worked as an Assistant Lecturer at the Electrical Engineering Department (EED) at Minia University. In November 2014, he joined the Electric Drives Laboratory (EDLAB) at the University of Padova in Italy for his Ph.D. research activities. Since May 2018, he has worked as an Assistant Professor at the Electrical Engineering Department at Minia University. He occupied a Postdoctoral Fellow position at the Department of Industrial Engineering, University of Padova, during the academic year 2021/2022. Currently, he is an Associate Professor in the EED at Minia University (Egypt). He is the author of two books and an editor of four books in international publication societies like Springer, Taylor&Francis and IntechOpen. He also served as a guest editor in international journals like *Energies and Machines*. He is an associate editor in the International *Journal of Robotics and Control Systems*, the *Discover Electronics Journal* at Springer and *Smart Grids and the Sustainable Energy Journal* at Springer. He also served as technical session chair at several international conferences. His research interests include renewable energy systems, power management, optimization, electric machine drives, power electronics, and load frequency control.

Mohit Bajaj is an Assistant Professor (Research) in the Department of Electrical Engineering at Graphic Era (Deemed to be University) in Dehradun, India. With a Ph.D. in Electrical Engineering from the prestigious National Institute of Technology in Delhi, he has established himself as a distinguished scholar and researcher in his field. Dr. Bajaj's academic journey includes a master's degree in Electrical Engineering from Motilal Nehru National Institute of Technology in Allahabad, India, specializing in Power Electronics and ASIC Design. He began his educational pursuit with a Bachelor of Technology in Core Electrical Engineering at Gurukula Kangri Vishwavidyalaya in Haridwar, India, an esteemed institution known for its rich heritage and academic excellence. Fueled by a passion for innovation and sustainability, Dr. Bajaj's primary research interests revolve around electric vehicles, renewable energy sources, distributed generation, power quality, and smart grids. His extensive contributions to the field are evidenced by his 250-plus research publications, comprising impactful journal articles, international conference papers, and book chapters. He has published over 150 research articles in SCI/SCIE indexed journals of reputed publishers such as IEEE, Elsevier, Wiley, Taylor and Francis, Springer, etc.

Contents

Preface

The global decarbonization goal is very ambitious, and the quick rise in renewable energy sources creates significant operational and design issues for the energy movement towards a target of 100% renewable energy. Power system operators are used to preserve system dependability in the face of past difficulties with unpredictable and fluctuating demand as well as planned and unforeseen outages. Concerns about the possible need for more flexibility in power networks are being raised by the additional difficulties brought about by our growing reliance on renewable energy sources (i.e. solar, wind, bioenergy, etc.). Energy storage technology options have been raised to be promising and flexible choices for maintaining highly reliable power systems.

This book aims to give readers a clear source of information about the value of energy storage technologies and the methodologies employed to evaluate that value. Representing the full potential of energy storage technology is becoming more and more crucial as utilities start incorporating flexibility into their resource planning procedures. This book is intended to serve as a summary and guide to the vast amount of pertinent research on energy storage that appears to be expanding exponentially. Additionally, the book presents sound research on energy storage materials. These materials are vital from a scientific study perspective, especially for energy storage applications. The high energy density ratios of energy storage materials make them useful for various applications. High-energy-density materials have a great storage capacity, which enables compact and efficient energy solutions.

Scientists and engineers interested in the latest advancements in energy storage systems are the target audience for this book. Particular attention is paid to energy professionals, including researchers, system operators, project managers, engineers and investors in energy storage projects.

Six chapters are offered inside 3 sections. Chapter 1, entitled "Hybrid and Advanced Energy Storage Systems: Integration, Applications, and Future Trends", concerns exploring hybrid energy storage systems such as battery-supercapacitor hybrids, thermal and electrical storage systems integration, and advancements in highperformance supercapacitors. It highlights key configurations, control strategies, and real-world applications for these technologies, including electric vehicles, renewable energy integration, and Internet of Things (IoT) systems. The chapter also investigates the challenges that still exist in the form of the requirement for novel electrode materials, such as silicon-based, nanocarbon-based, etc., implementation of novel coating methodologies such as 3D printing and inkjet printing to reduce the thickness, and so on; to promote utilization of supercapacitors on a commercial scale.

Following that, chapter 2 that is entitled "Optimizing Energy Storage Solutions for Grid Resilience: A Comprehensive Overview", presents a comprehensive review that examines the maturity, current status, and future directions of various storage technologies, with a primary focus on electrical, electrochemical, and thermal storage systems. The key findings highlight the advancements and potential of energy storage devices (ESDs). Some of the findings reported from the chapter are: i. Considerations of source availability, accessibility, and environmental impact are crucial for the life cycle analysis of electrochemical energy storage systems. ii. The availability of lithium and cobalt poses challenges to the current technology of Li-ion cells in terms of cost and scalability. iii. Utilizing thermal energy storage (TES) from renewable sources offers substantial potential to decrease CO_2 emissions across residential, non-residential, and industrial sectors by conserving significant amounts of energy.

Chapter 3 comes with the title of "Techno-Economic Assessment of a Grid-Connected Residential Rooftop Photovoltaic System with Battery Energy Storage System" and is concerned with investigating the feasibility of integrating six BESSs in a grid-connected solar PV installation for a residence situated in the North-Western part of Romania. The obtained results reveal that all investigated BESSs integrated with the PV systems are viable, but the PV installation combined with the 16.8 kWh Lead-acid battery is the most feasible variant for the examined dwelling. Although with Li-ion batteries, the proportion of green energy produced and consumed directly in the residence is higher than with Lead-acid batteries and, at the same time, investigated Li-ion batteries cover more of the home's consumption compared to Lead-acid batteries, the most viable option is a Lead-acid battery, mainly because Li-ion batteries require large capital expenditures. The attained results may assist Romanian stakeholders in acknowledging the economic feasibility of grid-connected PV systems combined with different BESSs technologies.

Under the title of "Electrochemical Performance and Design Strategies of MXene-Based Supercapacitors", chapter 4 explores MXene-based supercapacitors, detailing synthesis methods, electrochemical performance, and potential applications. MXene-based supercapacitor performance, covering specific capacitance, energy density, power density, and cycling stability are discussed. The chapter also addresses challenges in MXene-based supercapacitor development, including scalability, cost-effectiveness, and long-term stability, as well as applications in portable electronics, wearables, and electric transportation. Providing a comprehensive overview, this chapter inspires further exploration and utilization of MXene-based supercapacitors for next-gen energy storage solutions.

Chapter 5 is titled "Thermal Performance Analysis of RT 27 as PCM in Double-Pipe Heat Exchangers: Concentric and Hairpin Heat Exchangers". This chapter presents a study that aims to design latent heat energy storage systems for domestic and industrial applications. The study further reveals that the melting/solidification time depends on various parameters, including thermal diffusivity and viscosity of the PCM. Furthermore, the study emphasizes the importance of PCM selection, highlighting that PCMs with higher latent heat values can store more energy, but the rate of energy storage depends on the temperature difference between the high-temperature fluid and the initial temperature of the PCM. Overall, the findings suggest that through effective numerical analysis and optimization of design parameters, it is possible to propose energy storage systems that maximize efficiency and capacity, thereby contributing to the reduction of conventional fuel consumption and environmental impact.

With the title of “Revolutionizing Energy Applications: The Power of Interconnected Pores in Hierarchically Porous Carbon”, chapter 6 summarizes the design and synthesis of Hierarchically Porous Carbon (HPC) materials using hard-templating, soft-templating, and non-templating routes, with a focus on non-templating strategies for biopolymers. It discusses the recent use of HPCs and their composites in various electrochemical energy storage applications, such as supercapacitors, lithium-ion batteries, sodium-ion batteries, post-lithium-ion batteries, and hybrid energy storage devices. Moreover, the chapter offers insights into future challenges and research opportunities in HPC materials.

We want to express our gratitude to the authors, reviewers, and editorial staff for their contributions to this book.

Almoataz Y. Abdelaziz
Faculty of Engineering,
Ain Shams University,
Cairo, Egypt

Mahmoud A. Mossa
Faculty of Engineering,
Electrical Engineering Department,
Minia University,
Minia, Egypt

Mohit Bajaj
Graphic Era Hill University,
Dehradun, India

Section 1

Future Directions of Various Energy Storage Technologies

Chapter 1

Hybrid and Advanced Energy Storage Systems: Integration, Applications, and Future Trends

Khairy Sayed, Mahmoud Aref, Ahmed G. Abokhalil and Mahmoud A. Mossa

Abstract

Advanced and hybrid energy storage technologies offer a revolutionary way to address the problems with contemporary energy applications. Flexible, scalable, and effective energy storage is provided via thermal-electric systems, battery-supercapacitor hybrids, and high-performance supercapacitors. These technologies provide a sustainable route to the energy future and are essential to smart infrastructure, IoT systems, electric cars, and the integration of renewable energy. This chapter discusses how supercapacitors and battery systems can be combined to work better with vehicles' irregular energy needs. This is because the kinetics and thermodynamics of electrochemical reactions in battery technologies are not fast enough to meet those needs. Mini-grids, trains, trams, trucks, large off-road vehicles, tiny, uninterruptible power sources for Internet of Things nodes, and 1 MWh giants for hospitals and data centers are just a few of the applications that improved supercapacitors, and their derivatives offer enormous potential.

Keywords: energy storage, battery, Supercapacitor, energy future, IOT, mini-grids

1. Introduction

With the increasing demand for efficient, high-performance energy storage systems, hybrid and advanced energy storage systems have emerged as critical solutions for applications ranging from electric vehicles to smart grids. Combining the strengths of batteries, supercapacitors, and thermal energy storage technologies allows these systems to deliver both high and high energy density, enabling flexible and scalable storage solutions. This chapter explores hybrid energy storage systems such as battery-supercapacitor hybrids, thermal and electrical storage systems integration, and advancements in high-performance supercapacitors. It highlights key configurations, control strategies, and real-world applications for these technologies, including electric vehicles, renewable energy integration, and Internet of Things (IoT) systems [1–10].

Batteries are used as energy storage devices in multiple applications. However, batteries have certain disadvantages, such as low power density, limited life cycle, and comparatively slow response in certain applications. The feasibility of using a battery may be limited when dealing with transient high-power demands, such as the output power and load variability from transient renewable energy sources (RES). This may lead to an oversized design of batteries, resulting in increased investment cost and additional power loss due to the slow response of the batteries while compensating for transient peak power demands [11–15].

Electric vehicle (EV) batteries are particularly prone to degradation due to high peak power and harsh charging/discharging cycles during acceleration and deceleration. Moreover, battery-powered electric vehicle (BEV) applications require high power, resulting in an oversized battery pack and less optimal use of energy [16–20].

Supercapacitors (SC), on the other hand, do not store as much energy as batteries; and have the ability to accumulate and release the energy very rapidly. These devices are suitable for high-power vehicle applications, for providing power that is required to accelerate the vehicle or recover the available energy during braking phase. Supercapacitors cannot be used as the sole power source for EVs, as they have low energy density compared to batteries. However, they provide good options to compensate for the high peak of usage during short periods of time when battery power is not sufficient [21–27].

Supercapacitors, also known as electrostatic double-layer capacitors (EDLCs), are advanced energy storage devices that excel in rapid energy delivery and absorption. Their unique construction allows them to store energy through the electrostatic separation of charges, making them particularly effective in applications requiring quick bursts of power. For instance, they are commonly used in portable devices such as speakers and camera flashes, where fast recharge times are crucial. Additionally, in the realm of renewable energy, supercapacitors play a vital role in stabilizing the grid by managing fluctuations in power supply from sources like solar and wind. Their ability to recover and store energy during peak generation times enhances the efficiency of hybrid energy storage systems, contributing to a more sustainable energy landscape [28–33].

The advantages of supercapacitors over traditional batteries are significant, particularly in terms of performance and longevity. They can endure rapid charge and discharge cycles, making them ideal for applications that require frequent energy bursts, such as regenerative braking systems in transportation [34–40]. With a power density that is approximately a thousand times greater than that of batteries, supercapacitors can deliver high bursts of energy without the degradation associated with conventional battery technology. Moreover, their remarkable cycle stability allows them to maintain performance over more than one million charge-discharge cycles, far exceeding the lifespan of typical batteries. Supercapacitors also exhibit excellent temperature tolerance, enabling them to function effectively in extreme environments. Their unique construction, utilizing carbon-coated plates with minimal spacing, results in a larger electric field that enhances their energy storage capacity, positioning supercapacitors as a critical component in modern energy systems [41–48].

In modern energy systems, the integration of energy storage solutions plays a crucial role in enhancing the efficiency and reliability of microgrids [49–54]. **Figure 1** illustrates a DC microgrid integrated with energy storage, showcasing how direct current (DC) systems can effectively manage renewable energy sources such as solar panels. This configuration allows for seamless energy flow and storage, enabling optimized power distribution and utilization. Conversely, **Figure 2** depicts a hybrid AC/DC microgrid that incorporates energy storage, highlighting the versatility of combining alternating current (AC) and DC systems [55–59]. This hybrid approach not only facilitates the integration of diverse energy sources but also enhances grid stability and resilience by allowing for bidirectional power flow and improved load management. Together, these figures exemplify the innovative strategies employed to

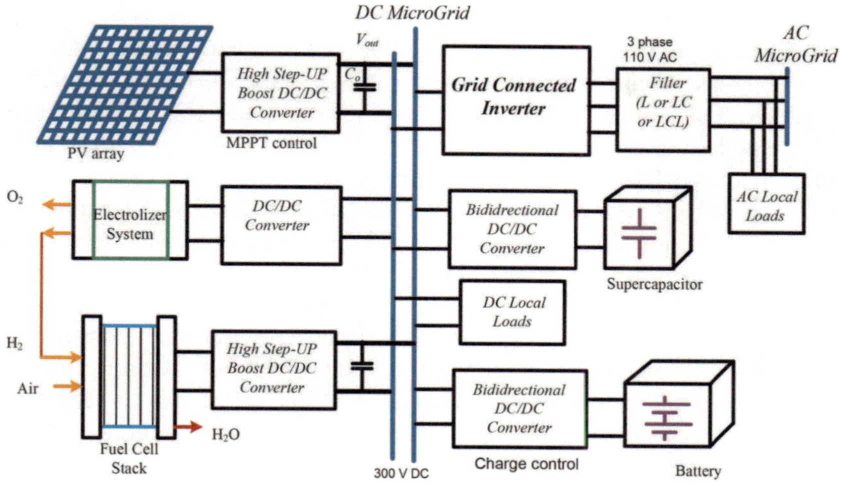

Figure 1.
DC microgrid integrated energy storage.

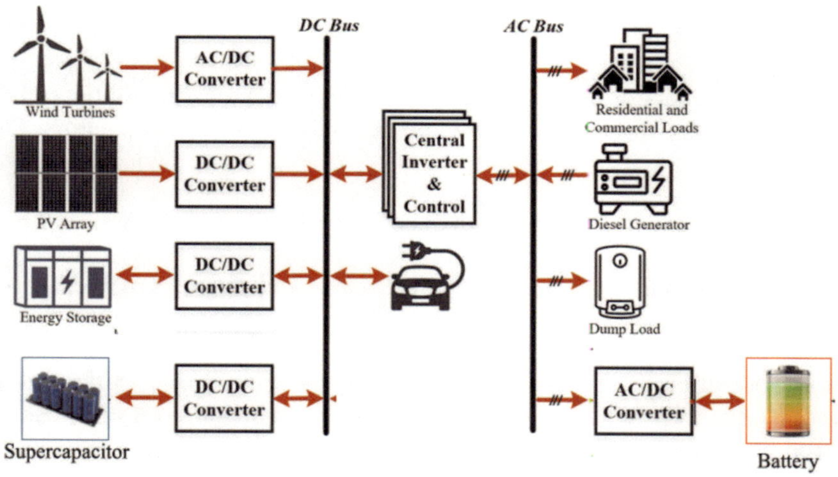

Figure 2.
Hybrid AC/DC microgrid integrated energy storage.

harness energy storage in microgrid applications, paving the way for more sustainable and adaptable energy solutions.

Battery-supercapacitor hybridization helps overcome the limitations of batteries or supercapacitors. It reduces the stresses applied to batteries, thus improving their life [4].

The hybridization of the embedded energy storage systems provides the following advantages [4, 5]:

- Improved Li-ion battery lifetime
- Maximized energy recovery during braking
- Reduced size of embedded energy storage system
- Reduced cost of embedded source

Electrification is an important means of decreasing greenhouse gas emissions in the transportation sector. The global electric car fleet has now exceeded 5 million and will continue to increase in the future. The energy storage system is a critical part of the electric vehicle. The storage system has to be cost-effective, light, efficient, safe, reliable, occupy less space, and have a long life. It should also be produced and disposed of in an eco-friendly way. Interestingly, electric vehicles can be used as backup storage during periods of grid failure or spikes in demand. Elverlingsen in Germany collects almost 2000 batteries from Mercedes Benz EVs to create a stationary grid-sized battery that can hold 9 MW of energy [60–64].

The lithium-ion (Li-ion) battery technology plays a major role in EVs due to its power and energy density. Supercapacitors (SC) possess extremely high power density, high cycle time, cycling efficiency, and low energy density. Connecting battery packs with supercapacitors is considered to be an effective method to provide energy and power for EVs and Hybrid Electric Vehicles (HEVs) as it results in both high power and high energy capability [65–73].

2. Battery-supercapacitor hybrids for multi-functional applications

2.1 Overview of battery-supercapacitor hybrids

Battery-supercapacitor hybrids combine the energy storage capabilities of batteries with the rapid charge-discharge and high power delivery of supercapacitors. These systems integrate:

- *Battery-type electrode*: Provides high energy density for sustained energy supply.
- *Capacitive electrode*: Delivers rapid power bursts and supports high charge-discharge rates.

This hybridization offers a balance of *high energy density*, *long cycle life*, and *rapid response times*, making them ideal for applications that demand both sustained energy supply and peak power delivery.

Table 1 shows a comprehensive comparison table that outlines the characteristics of batteries, supercapacitors, battery-supercapacitor hybrids, thermal and electrical storage systems integration, and advanced high-performance supercapacitors.

Batteries are renowned for their high energy density, making them essential for long-duration energy applications such as electric vehicles (EVs) and grid storage, though they face challenges with cycle life and power density. In contrast, supercapacitors excel in scenarios requiring rapid power delivery, ultra-fast charging, and extended lifespans but are limited by their lower energy density. To address these trade-offs, battery-supercapacitor hybrids integrate the strengths of both technologies, creating versatile solutions ideal for EVs and renewable energy systems. Additionally, integrating thermal and electrical storage is increasingly vital for industrial and renewable applications where managing both heat and electricity is crucial. Emerging advancements in high-performance supercapacitors, driven by innovative materials like MXenes and metal-organic frameworks (MOFs), hold the potential to narrow the energy density gap between supercapacitors and batteries, broadening their applicability.

DOI: http://dx.doi.org/10.5772/intechopen.1010067

Feature	Battery	Supercapacitor	Battery-supercapacitor hybrids	Thermal & electrical storage systems	Advanced high-performance supercapacitors
Energy density	High (100–265 Wh/kg for Li-ion)	Low (5–10 Wh/kg)	Moderate (higher than supercapacitors but less than batteries)	High for thermal (depending on PCM/molten salt) and moderate for electrical	Improved with nanomaterials like MXenes and MOFs, achieving 20–50 Wh/kg
Power density	Moderate (200–300 W/kg)	Very high (>10,000 W/kg)	High (enhanced by supercapacitor component)	Thermal: low; electrical: high	High (due to advanced materials enabling faster charge/discharge)
Charge/discharge time	Minutes to hours	Seconds	Fast charging for high-power demands, slower for energy storage	Thermal: slow (hours); electrical: fast (seconds to minutes)	Rapid (seconds to minutes)
Cycle life	Limited (~500–3000 cycles for Li-ion)	Extremely high (>1,000,000 cycles)	Improved compared to standalone batteries	Depends on system design (thermal materials degrade over time)	Extremely high (similar to supercapacitors)
Applications	Energy storage for EVs, portable devices, grid	High-power applications (e.g., regenerative braking, and UPS)	EVs, HEVs, IoT systems, renewable energy integration	Industrial heat recovery, renewable energy systems	EVs, renewable energy storage, IoT, and wearables
Cost	Moderate to high	Low to moderate	Moderate to high	Moderate (low cost for thermal, high for electrical)	Reducing with scalable production of advanced materials
Efficiency	Moderate (~85–90%)	Very high (>95%)	High (combines strengths of batteries and supercapacitors)	High for combined systems when well-designed	Very high (due to reduced resistive losses and material innovations)
Environmental impact	Resource-intensive (e.g., lithium and cobalt mining)	Low environmental impact	Moderate (hybrids reduce battery reliance)	Depends on materials (thermal often eco-friendly)	Potentially low (using sustainable materials like MOFs)
Key advantages	High energy density	High power density and long lifespan	Combines high energy and power density; extends battery life	Enables both heat and electricity storage for integrated systems	Improved performance, higher operating voltage, lightweight and compact designs
Key limitations	Limited lifespan, slower charge/discharge	Low energy density	More complex integration and costlier than individual components	Complex integration and thermal losses	High material costs (MXenes, MOFs), scalability challenges
Emerging trends	Solid-state batteries, recycling improvements	Hybrid SCs with higher energy density	Advanced energy management systems for better synergy	Thermo-electrochemical integration for efficiency	Use of advanced nanomaterials (e.g., MXenes and graphene) to improve performance

Table 1.
Comprehensive comparison of ESSs.

3. Applications of battery-supercapacitor hybrids

Figure 3 illustrates the diverse applications of hybrid energy storage systems (HESS), showcasing their versatility across multiple sectors. In renewable energy integration, HESS plays a crucial role in stabilizing fluctuations from sources like solar and wind, ensuring a consistent power supply. In electric vehicles, these systems enhance performance by optimizing energy usage and extending battery life. Additionally, HESS finds applications in grid support, where it aids in load leveling and frequency regulation, contributing to a more resilient energy infrastructure. Other notable uses include uninterruptible power supplies (UPS) for critical facilities and smart grids, highlighting the growing importance of HESS in modern energy solutions.

3.1 Electric vehicles (EVs) and hybrid electric vehicles (HEVs)

Battery-supercapacitor hybrids significantly enhance the performance and efficiency of electric vehicles (EVs) and hybrid electric vehicles (HEVs). One of the key benefits is the implementation of regenerative braking, where supercapacitors capture energy during braking and store it for use during acceleration. This process not only improves overall energy efficiency but also contributes to a smoother driving experience. Furthermore, supercapacitors are adept at handling short, high-power demands, which reduces the strain on the vehicle's battery. This capability not only extends the lifespan of the battery but also leads to better fuel efficiency. In the case of HEVs, the reduced fuel consumption directly correlates with lower emissions, making these vehicles more environmentally friendly.

3.2 Smart meters

Hybrid supercapacitors play a crucial role in powering wireless communication links within smart meters. Their ability to provide reliable operation during data transmission and power outages ensures that these devices function effectively, even in challenging conditions. By maintaining consistent communication, hybrid

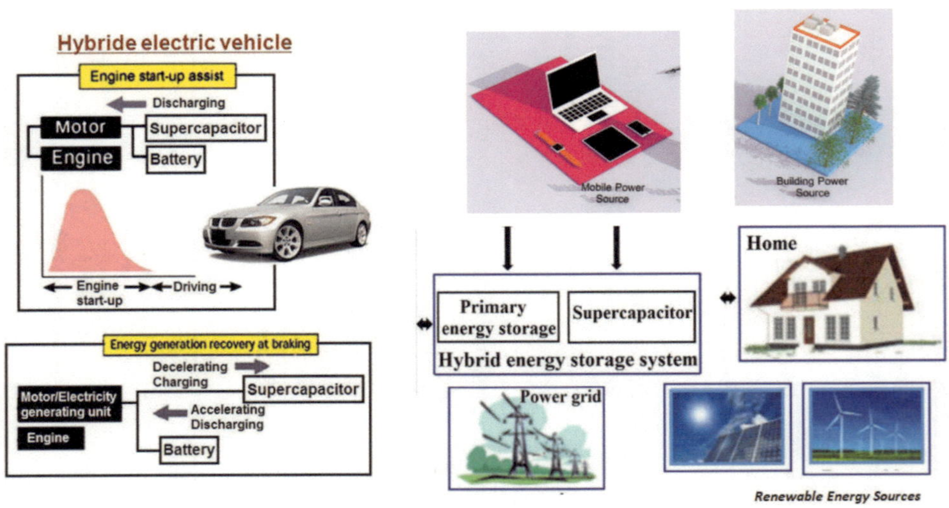

Figure 3.
Various applications of HESS.

supercapacitors enhance the overall reliability of smart grid systems, which are essential for modern energy management.

3.3 Industrial processes and Programmable Logic Controllers (PLCs)

In industrial environments, hybrid supercapacitors are instrumental in providing "ride-through" power during brief power outages. This capability is vital for supporting operations and preventing interruptions in processes controlled by Programmable Logic Controllers (PLCs). By ensuring a continuous power supply during transient disturbances, hybrid supercapacitors help maintain productivity and minimize downtime in manufacturing and other industrial applications.

3.4 Data centers

Data centers rely heavily on hybrid supercapacitors as backup power sources for critical components such as volatile cache memory, servers, and RAID storage systems. These supercapacitors ensure data integrity by providing instantaneous power during grid failures, thereby preventing service interruptions. The ability to quickly deliver power is essential for maintaining operational continuity in data centers, where even brief outages can lead to significant disruptions and data loss.

3.5 Internet of Things (IoT) systems

Battery-supercapacitor hybrids are particularly well-suited for Internet of Things (IoT) devices that require efficient energy storage and rapid power delivery. Their lightweight design allows for easy integration into low-power IoT networks, while their fast charging capabilities ensure that devices remain operational without long downtimes. Additionally, the long lifecycle of hybrid supercapacitors makes them an ideal choice for IoT applications, where longevity and reliability are paramount.

3.6 Renewable energy sources

In the realm of renewable energy, battery-supercapacitor hybrids serve multiple functions. They help smooth out power fluctuations associated with intermittent energy sources like solar and wind, stabilizing the output and making it more reliable for grid integration. Furthermore, these hybrids can store excess energy generated during peak production times and deliver it during periods of low output, enhancing overall grid support and promoting a more sustainable energy system.

Hybrid supercapacitors integrate the features of both batteries and supercapacitors into a single unit, transcending the mere combination of these two technologies in one casing. They represent energy sources that blend battery chemistry with supercapacitor physics, effectively addressing the limitations of each while offering distinct advantages for developers in achieving design goals. For IoT system designers, hybrid supercapacitors offer a compelling option for energy storage and power delivery due to their high energy densities, extended cycle lifetimes, and elevated working voltages. Utilizing these components allows for fewer cells and reduced volume compared to traditional supercapacitors, while also better satisfying temperature and longevity requirements than batteries alone. By eliminating challenging trade-offs, hybrid supercapacitors empower design engineers to meet demanding project specifications more effectively.

4. Configurations of HESS in microgrid

Hybrid energy storage systems (HESS) in microgrids combine different energy storage technologies, such as batteries and supercapacitors, to optimize performance by leveraging their complementary characteristics. **Figure 4** shows the classification of the battery-supercapacitor HESS topologies.

Figure 4, which illustrates the classification of battery-supercapacitor hybrid energy storage systems (HESS) topologies, it would be essential to consider the general structure and categories typically represented in such diagrams. The classification likely divides HESS into several main categories based on their configuration and operational principles. Common classifications include series, parallel, and series-parallel configurations. Each topology has distinct advantages and limitations concerning energy density, power density, efficiency, and control complexity.

Series configuration: In a series topology, the battery and supercapacitor are connected. This configuration can be beneficial for applications requiring high-voltage output. However, it may limit the overall energy capacity to that of the weakest component, which can be a drawback in certain scenarios.

Parallel configuration: In a parallel topology, both the battery and supercapacitor are connected in parallel, allowing them to share the load. This setup can enhance power delivery and improve efficiency but might complicate the management of charge and discharge cycles between the two components.

Hybrid configuration: Some systems may employ a hybrid approach, combining elements of both series and parallel configurations. This allows for a more tailored approach, optimizing performance based on specific application requirements.

Figure 5 shows different configurations of HESS. These configurations are designed to balance energy and power demands efficiently. The key configurations include:

1. Parallel active HESS topology:

    - Both energy storage devices (e.g., batteries and supercapacitors) are connected to a common DC bus through individual power converters.

    - Each device operates independently, allowing optimized power management and the ability to handle different power and energy demands simultaneously.

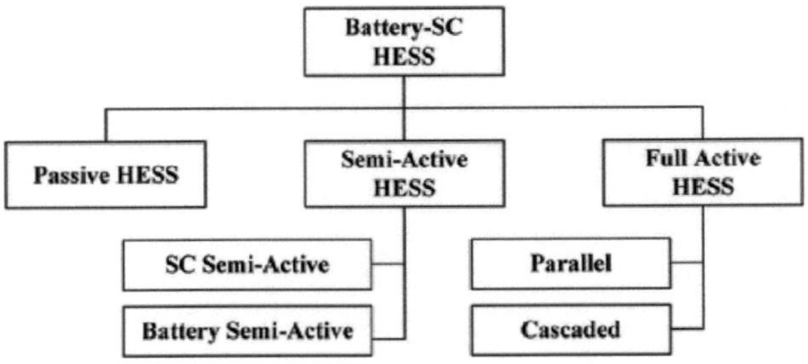

Figure 4.
Classification of the battery-supercapacitor HESS topologies.

DOI: http://dx.doi.org/10.5772/intechopen.1010067

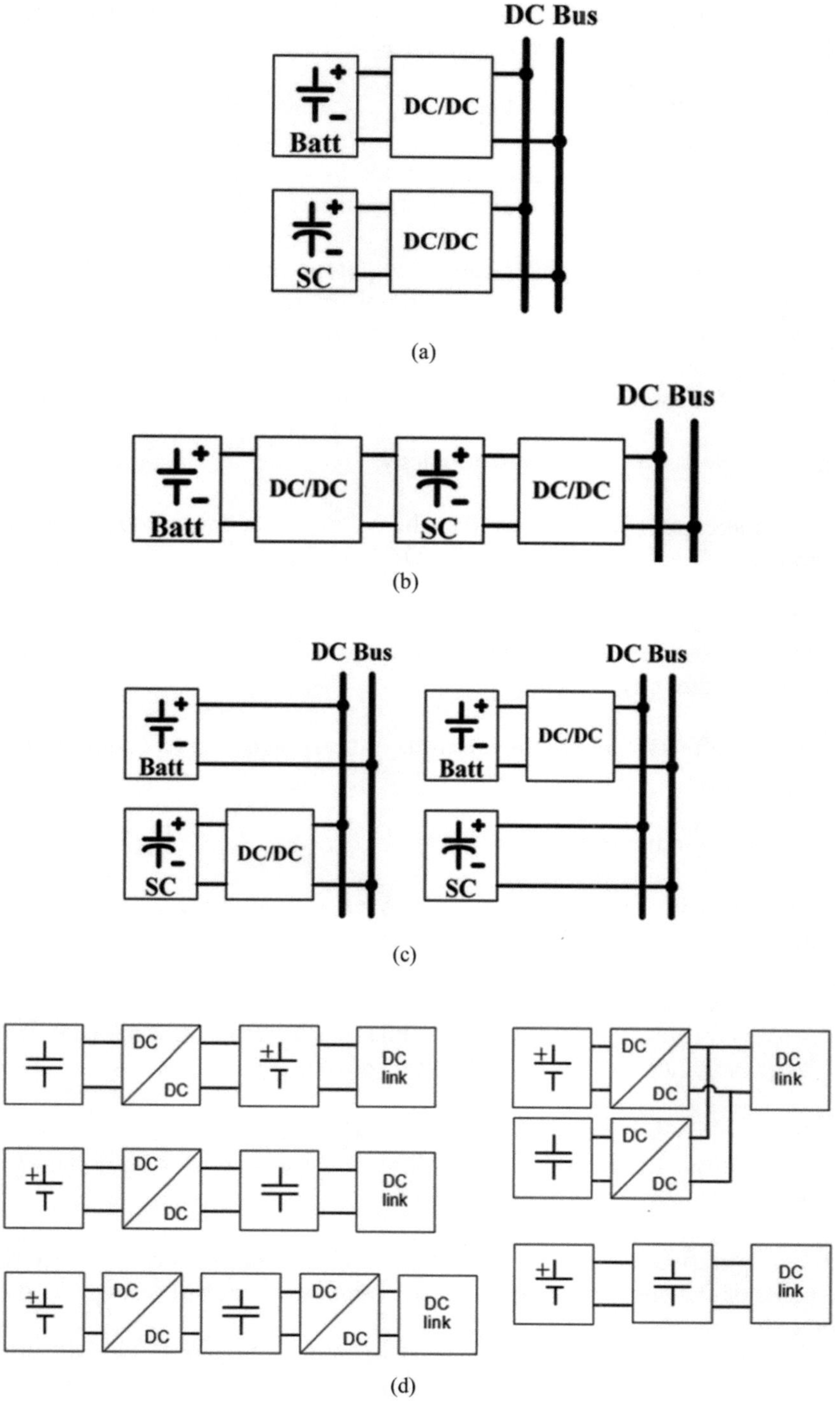

Figure 5.
HESS topologies: (a) parallel active HESS topology, (b) cascaded active HESS topology, (c) supercapacitor semi-active HESS topology, and (d) battery semi-active HESS topology.

- This configuration offers high flexibility, enhanced dynamic performance, and better utilization of both storage devices.

2. Cascaded ACTIVE HESS topology:

   - The energy storage devices are connected in series, with one device directly interfaced with the DC bus and the other interfaced through a converter.
   - This configuration simplifies control requirements and reduces the number of converters compared to the parallel topology, but it can introduce some design challenges related to managing the cascaded connections.
   - It is suitable for systems where reduced hardware complexity is desired.

3. Supercapacitor semi-active hess topology:

   - The supercapacitor is connected to the DC bus *via* a power converter, while the battery is directly connected to the DC bus.
   - This topology allows the supercapacitor to handle high power transients and rapid changes in load demand, protecting the battery from excessive cycling and prolonging its lifespan.
   - It is cost-effective and widely used in systems requiring frequent peak power delivery.

4. Battery semi-active HESS topology:

   - The battery is connected to the DC bus *via* a power converter, while the supercapacitor is directly connected to the DC bus.
   - In this configuration, the battery operates under controlled conditions, focusing on steady energy supply, while the supercapacitor manages transient power demands.
   - This topology is particularly effective in applications where energy density is more critical, and the supercapacitor serves as a buffer for rapid power changes.

Each configuration has distinct advantages and is chosen based on the specific requirements of the microgrid, such as cost, complexity, control strategy, and the need for energy or power optimization.

Topology options come with trade-offs. Due to the inherent design and performance differences between batteries and electric double-layer capacitors (EDLCs), designers must choose whether to utilize a single energy storage device or a combination of both. If they opt for a hybrid approach, they need to consider various topologies, each presenting its own trade-offs and implications for performance (see **Figure 6**). **Figure 6** illustrates three common ways to combine a supercapacitor and a battery: in parallel, as independent units, or integrated through a controller/regulator.

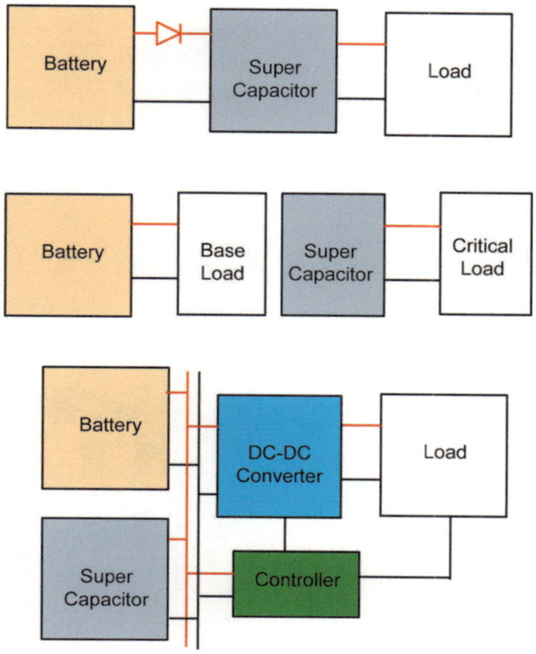

Figure 6.
Topology's own trade-offs and implications for performance.

- The parallel configuration is the simplest option; however, it does not fully optimize the supercapacitor's capabilities, and its output voltage is directly linked to the battery voltage.

- Utilizing the battery and supercapacitor as independent units is most effective when there is a non-critical base load alongside a separate critical load, as it allows for independent power supply to each. However, this method does not leverage any synergy between the two units.

- The smart arrangement integrates the strengths of both energy sources, maximizing both runtime and cycle life. This setup requires additional management components like a controller and DC-DC regulation between the energy sources and the load, making it particularly suitable for transportation-related power systems.

When considering these topologies, selecting a battery and a supercapacitor is not simply an "either/or" choice. Designers can choose to implement both, but doing so necessitates finding the right balance between their differing characteristics. Fortunately, thanks to an innovative component, there is no longer a need to choose between batteries, supercapacitors, or both. Eaton – Electronics Division offers a range of hybrid energy-storage components that merge the benefits of both technologies into a single package, eliminating the need for compromise.

Figure 7 illustrates two distinct microgrid configurations—AC-coupled and DC-coupled—integrated with supercapacitors, showcasing the essential role of both AC/DC and DC-DC converters in these setups. In the AC-coupled configuration, supercapacitors are connected to the AC bus, allowing for seamless energy storage and

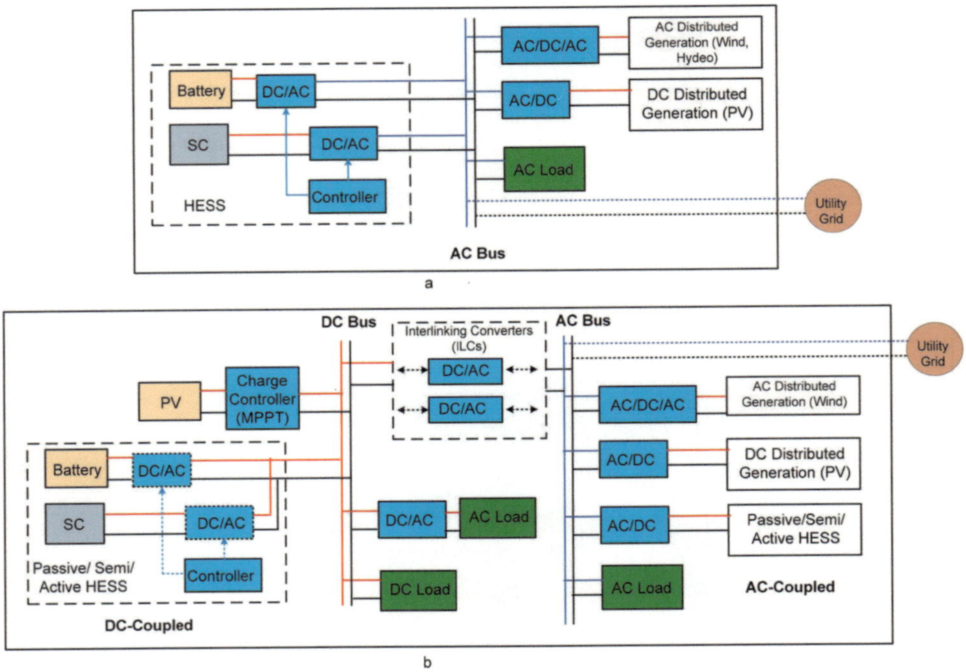

Figure 7.
AC coupled and Dc coupled microgrid configurations interfaced supercapacitors including AC/DC and DC-Dc converters.

discharge while facilitating interaction with the grid and other AC sources. This configuration typically employs AC/DC converters to manage the conversion of energy between AC and the supercapacitor's storage system. Conversely, the DC-coupled configuration connects supercapacitors directly to the DC bus, enabling more efficient energy transfer with reduced conversion losses. In this setup, DC-DC converters are utilized to regulate voltage levels and optimize power flow between the supercapacitors and other renewable energy sources, such as solar panels or wind turbines. Both configurations highlight the versatility of supercapacitors in enhancing energy management, improving system reliability, and supporting the integration of renewable energy sources within microgrid applications.

Overall, **Figure 7** serves as a valuable resource for understanding the diverse approaches to integrating batteries and supercapacitors within hybrid energy storage systems. By classifying these topologies, it aids engineers and researchers in selecting the appropriate configuration for their specific needs and applications.

5. Battery-supercapacitor hybrid energy storage systems in electric vehicles

Battery-supercapacitor hybrid energy storage systems (HESS) are increasingly utilized in electric vehicles (EVs) to optimize performance by combining the high energy density of batteries with the high power density and fast charge/discharge capabilities of supercapacitors. Different configurations of battery-supercapacitor HESS are designed to meet the specific power and energy requirements of EVs. These configurations include:

5.1 Passive parallel configuration

- *Description*: The battery and supercapacitor are directly connected in parallel to the DC bus without any additional power converters.
- *Operation*: The supercapacitor handles high-frequency power demands (e.g., acceleration and regenerative braking), while the battery supplies steady energy for driving.
- *Advantages*:
    - Simple and cost-effective design.
    - Minimal control requirements.
- *Disadvantages*:
    - Limited control over the power-sharing between the battery and supercapacitor, potentially leading to inefficient utilization.

5.2 Semi-active configuration

- Battery semi-active configuration:
    - *Description*: The battery is connected to the DC bus *via* a bidirectional DC-DC converter, while the supercapacitor is directly connected to the DC bus.
    - *Operation*: The converter regulates the battery's power delivery, ensuring it primarily supplies steady-state energy, while the supercapacitor manages transient power demands.
    - *Advantages*:
        - Protects the battery from high-power cycling, extending its lifespan.
        - Enhanced power management for rapid changes.
    - *Disadvantages*:
        - Requires a converter, increasing cost and complexity.
- *Supercapacitor semi-active configuration*:
    - *Description*: The supercapacitor is connected to the DC bus *via* a DC-DC converter, while the battery is directly connected.
    - *Operation*: The converter controls the supercapacitor, allowing it to handle rapid power surges and regenerative braking.
    - *Advantages*:
        - Efficient energy recovery and fast power response.

- *Disadvantages*:
    - Slightly less efficient battery protection compared to the battery semi-active configuration.

Figure 8 depicts a fully active configuration of a microgrid, characterized by its comprehensive integration of various energy sources and storage systems to optimize energy management and reliability. In this configuration, all components, including renewable energy sources, energy storage systems, and loads, are actively controlled and monitored to ensure seamless operation. The system employs advanced control strategies that allow for real-time adjustments in response to changing demand and generation conditions. By utilizing smart inverters and sophisticated communication technologies, the fully active configuration facilitates efficient energy flow and enhances the stability of the microgrid. This setup not only maximizes the utilization of renewable energy but also improves the resilience of the grid against disturbances. Overall, **Figure 8** illustrates how a fully active microgrid configuration can effectively support sustainable energy goals while providing reliable power to consumers.

5.3 Fully active configuration

- *Description*: Both the battery and supercapacitor are connected to the DC bus through separate bidirectional DC-DC converters.
- *Operation*: The converters independently control the power delivery of both the battery and the supercapacitor, allowing for optimal energy and power management.
- *Advantages*:
    - Maximum flexibility and efficiency in power-sharing.
    - Extended battery life due to precise load management.
    - Improved performance during high-power events like acceleration and braking.

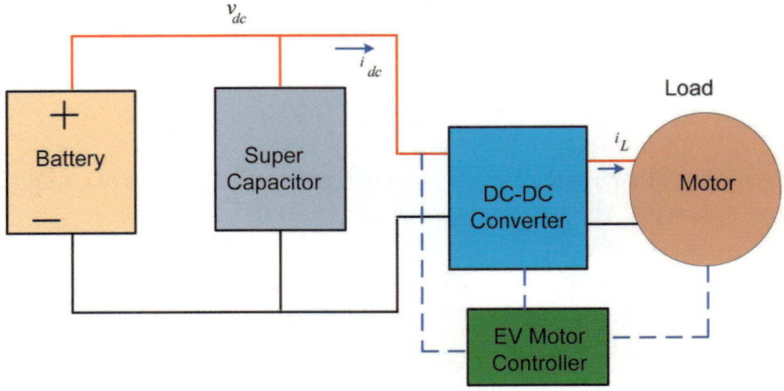

Figure 8.
Fully active configuration.

- *Disadvantages*:
  - High cost and complexity due to additional hardware and control requirements.

5.4 Cascaded configuration

- *Description*: The supercapacitor is connected to the DC bus *via* a converter, and the battery is connected to the DC bus indirectly through the supercapacitor.
- *Operation*: The supercapacitor acts as an intermediary, buffering high-frequency power demands and reducing stress on the battery.
- *Advantages*:
  - Protects the battery from transient loads.
  - Efficient energy buffering.
- *Disadvantages*:
  - More complex control and potential energy losses in cascading.

5.5 Application in electric vehicles

- *Acceleration and deceleration*: The supercapacitor manages the rapid power delivery during acceleration and absorbs energy during regenerative braking, reducing the battery's cycling load.
- *Battery protection*: By smoothing out high-frequency power demands, the battery operates under more stable conditions, enhancing its lifespan and reliability.
- *Efficiency and range*: Optimal energy and power management improve overall system efficiency, contributing to extended driving range and better performance.

The choice of configuration depends on factors like cost, complexity, and performance requirements of the specific EV application. Fully active systems, while complex, provide the best balance for high-performance EVs, whereas passive and semi-active systems are more suitable for cost-sensitive designs.

Figure 9 depicts the energy management control system for a hybrid energy storage system (HESS) in electric transportation, specifically highlighting the integration of a DC catenary line. This system is designed to optimize the flow of power between various energy sources, including the catenary line, batteries, and supercapacitors. The control mechanism monitors real-time energy demands and adjusts the distribution of power accordingly. When the vehicle is connected to the catenary line, it can draw energy directly to charge the battery or supply power to the traction system. During peak demand scenarios, such as acceleration or climbing, the system can seamlessly tap into the supercapacitor's rapid discharge capabilities to provide additional power, ensuring smooth operation and improved performance.

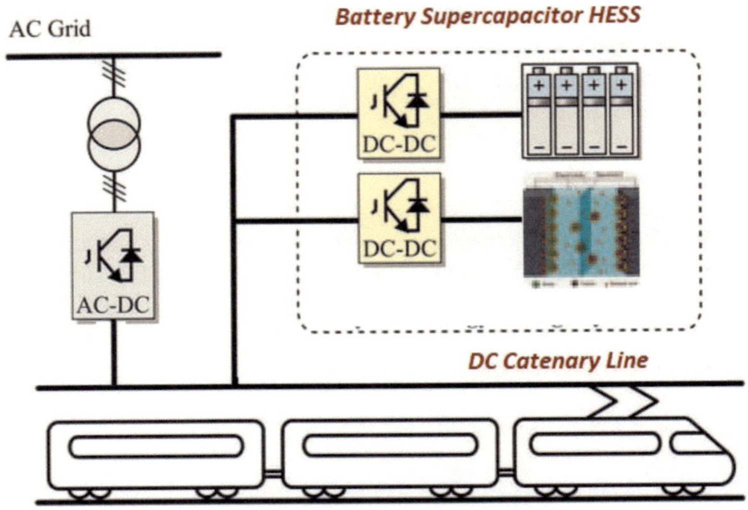

Figure 9.
HESS in electric transportation.

Conversely, during regenerative braking, the energy generated can be stored back into the supercapacitors or fed into the catenary line, enhancing energy efficiency. This intelligent management of power flow not only maximizes energy utilization but also contributes to reduced operational costs and improved sustainability in electric transportation systems.

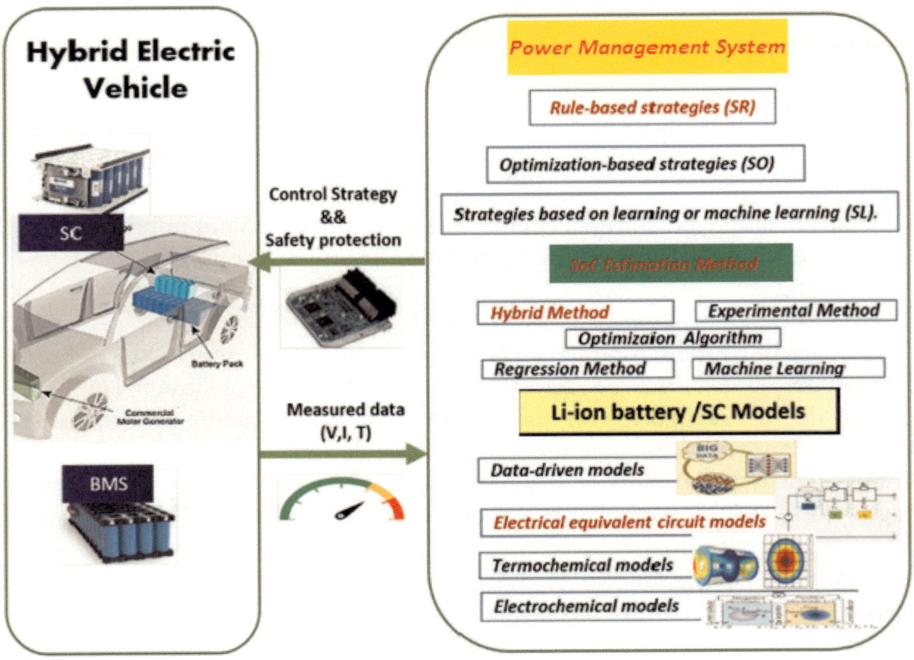

Figure 10.
Power management in EV.

Figure 10 illustrates the power management system in electric vehicles (EVs) that incorporates a hybrid energy storage solution combining supercapacitors and batteries. This integrated approach optimizes energy efficiency and performance by leveraging the rapid charge and discharge capabilities of supercapacitors alongside the higher energy density of batteries. In this configuration, the power management system intelligently coordinates the energy flow between the two storage devices, ensuring that the battery handles long-term energy supply while the supercapacitor addresses short bursts of high power demand, such as during acceleration or regenerative braking. By effectively managing the energy distribution, this hybrid setup enhances overall vehicle performance, extends battery life, and improves energy recovery during driving cycles, ultimately contributing to a more efficient and sustainable electric vehicle operation.

6. Control schemes of HESS

Control schemes for hybrid energy storage systems (HESS) are essential for managing the power flow between energy storage components, such as batteries and supercapacitors, to ensure optimal performance, efficiency, and longevity. These control schemes aim to balance power and energy demands while protecting the storage devices from excessive cycling and stress. Below are the key control schemes commonly used for HESS:

6.1 Rule-based control

- *Description*: Predefined rules determine the power-sharing between the battery and the supercapacitor based on operational requirements.
- *Mathematical form*:

$$\begin{gathered} \textit{If } P_{load} > P_{thershold} \textit{ and } SOC_{sc} > SOC_{sc,min} \\ \text{Then } P_{sc} = \min(P_{load}, P_{sc,max}) \\ \text{and } P_{batt} = P_{load} - P_{sc} \end{gathered}$$

where $P_{thershold}$: Thershold for high-power demand.
$SOC_{sc,min}$: Minimum SoC of the supercapacitor.
$P_{sc,max}$: Maximum power the supercapacitor can supply.

Types:

- *Threshold-based control*: Power is distributed based on a specific threshold, such as current or voltage levels.

  Mathematical form:

$$\begin{gathered} \textit{If } P_{load} \le P_{thershold} \\ \text{Then } P_{batt} = P_{load} \\ \text{and } P_{sc} = 0 \end{gathered}$$

- *State of charge (SoC) control*: Power allocation is based on the SoC of the battery and supercapacitor.

  Mathematical form:

$$\text{If } SOC_{batt} < SOC_{batt,min}$$
$$\text{Then } P_{sc} = P_{load}$$
$$\text{and } P_{batt} = 0$$

- *Advantages*:
  - Simple implementation with minimal computational requirements.
  - Reliable and effective for basic systems.
- *Disadvantages*:
  - Limited adaptability to dynamic operating conditions.
  - Suboptimal energy management in complex scenarios.

6.2 Frequency decoupling control

- *Description*: Power demand is divided into low-frequency (steady-state) and high-frequency (dynamic) components using filters.
  - The battery handles low-frequency components, supplying steady energy.
  - The supercapacitor addresses high-frequency components, managing rapid power transients.
- *Mathematical form:*

  Low-frequency component (P_{lf}):

$$P_{lf} = LPF(P_{load})$$

where LPF is a low-pass filter with a cutoff frequency (f_c) that separates low-frequency content

High-frequency component (P_{hf}):

$$P_{hf} = P_{load} - P_{lf}$$

Battery SoC:

$$SOC_{batt}(t+1) = SOC_{batt}(t) - \frac{P_{batt}.\Delta t}{E_{batt}}$$

where: E_{batt}: Total energy capacity of the battery
Δt: Time step.
Supercapacitor SoC:

$$SOC_{sc}(t+1) = SOC_{sc}(t) - \frac{P_{sc}.\Delta t}{E_{sc}}$$

where: E_{sc}: Total energy capacity of the Supercapacitor

- *Advantages*:
    - Effective separation of power demands reduces battery stress.
    - Enhances the system's dynamic response and lifespan.
- *Disadvantages*:
    - Requires accurate filter design and tuning.
    - May introduce delays in dynamic response.

Figure 11 illustrates various control schemes that can be implemented in hybrid energy storage systems (HESS). These include rule-based control, which uses predefined thresholds for power-sharing, and frequency decoupling control, which separates power demands into high- and low-frequency components to optimize device usage. Advanced methods like model predictive control (MPC) enable real-time optimization by predicting future power requirements, while fuzzy logic control provides flexibility in managing nonlinear and uncertain system dynamics. Additionally, droop control facilitates decentralized power-sharing based on system conditions, and AI-based control leverages machine learning to enhance adaptability and efficiency. Each

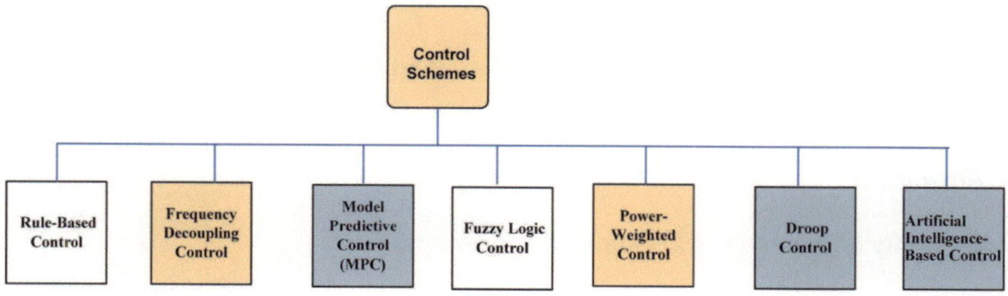

Figure 11.
Typical control schemes of HESS.

scheme offers unique advantages and is suited for specific HESS applications depending on complexity, performance requirements, and system dynamics.

6.3 Model predictive control (MPC)

- *Description*: MPC uses a mathematical model of the HESS to predict future power demands and optimizes power-sharing over a defined prediction horizon.
- *Operation*:
  - o Real-time optimization considers constraints such as SoC limits, power capacity, and efficiency.
  - o Ensures the system operates at maximum efficiency while protecting the storage devices.
- *Mathematical form*:

  Battery dynamics:

$$SOC_{batt}(k+1) = SOC_{batt}(k) - \frac{P_{batt}(k).\Delta t}{E_{batt}}$$

where: $SOC_{batt}(k)$: Battery SoC at time step k.
$P_{batt}(k)$: Battery power at time step k (positive for discharge, negative for charge)
E_{batt}: Total energy capacity of the battery
Δt: Time step

Supercapacitor dynamics:

$$SOC_{sc}(k+1) = SOC_{sc}(k) - \frac{P_{sc}(k).\Delta t}{E_{sc}}$$

where: $SOC_{sc}(k)$: Supercapacitor SoC at time step k.
$P_{sc}(k)$: Supercapacitor power at time step k (positive for discharge, negative for charge)
E_{sc}: Total energy capacity of the Supercapacitor
Δt: Time step

- *Advantages*:
  - o Highly adaptable to varying load conditions.
  - o Optimized performance with improved energy efficiency.
- *Disadvantages*:
  - o Computationally intensive, requiring advanced processors.
  - o Complexity in modeling and implementation.

6.4 Fuzzy logic control

- *Description*: A fuzzy logic controller uses linguistic rules to determine power-sharing between the battery and supercapacitor.
- *Operation*:
  - Inputs: Parameters such as load power, SoC, and power rate.
  - Outputs: Control signals to allocate power dynamically.

    Fuzzy rules are expressed in the form of "if-then" statements:

    1. If P_{load} is High and SOC_{sc} is High, then P_{sc} is High.
    2. If P_{load} is Medium and SOC_{sc} is Medium, then P_{sc} is Medium.
    3. If P_{load} is Low and SOC_{sc} is Low, then P_{sc} is Low.
- *Advantages*:
  - Handles nonlinear and uncertain systems effectively.
  - Does not require a detailed mathematical model.
- *Disadvantages*:
  - Performance depends heavily on the quality of rule design.
  - May require extensive tuning for optimal results.

6.5 Power-weighted control

- *Description*: Power-sharing is determined by weighting the contributions of the battery and supercapacitor based on their characteristics.
- *Operation*:
  - Batteries handle the bulk of energy demand, while supercapacitors manage peak power.
  - Weights are dynamically adjusted based on system conditions.
- *Advantages*:
  - Simple yet effective for real-time power-sharing.
  - Reduces stress on the battery by offloading transient power.
- *Disadvantages*:
  - Requires accurate weight calculation.
  - Less flexible in highly dynamic systems.

6.6 Droop control

- *Description*: Power-sharing is based on the droop characteristics of the battery and supercapacitor, which are adjusted according to their SoC or voltage.
- *Operation*:
  - o Devices with higher available capacity contribute more to the load.
  - o Power-sharing is proportional to the droop coefficients.
- *Mathematical form*:

  The droop equation defines the relationship between the output of a storage component and a system parameter (e.g., SoC, frequency, or voltage). For a HESS, the droop equation is expressed as:

$$P_i = P_{i,ref} - k_i.\left(SOC_i - SOC_{i,ref}\right)$$

where P_i: Power output of the ith storage component (battery or supercapacitor).
$P_{i,ref}$: Reference power output of the ith storage component.
k_i: Droop coefficient of the ith storage component.
SOC_i: state of charge of the ith storage component.
$SOC_{i,ref}$: Reference state of charge of the ith storage component

- *Advantages*:
  - o Enables decentralized control, reducing communication requirements.
  - o Ensures smooth transitions in power-sharing.
- *Disadvantages*:
  - o Performance depends on accurate droop coefficient design.
  - o Suboptimal for systems with high variability in load.

6.7 Artificial intelligence-based control

- *Description*: AI techniques, such as machine learning or neural networks, optimize power-sharing based on historical data and real-time inputs.
- *Operation*:
  - o Algorithms learn from system behavior to predict and allocate power dynamically.

The AI model can be a neural network, reinforcement learning agent or other machine learning model. The model is trained to minimize a cost function or maximize a reward function.

Neural Network Example:

$$\mathbf{y} = f_{NN}(\mathbf{x}; \mathbf{W}, \mathbf{b})$$

where $\mathbf{x}$: input vector (e.g., $[P_{load}, SOC_{batt}, SOC_{sc}, P_{gen}]$)
$\mathbf{y}$: output vector (e.g., $[P_{batt}, P_{sc}]$)
f_{NN}: Neural network function with weights $\mathbf{W}$ and biases $\mathbf{b}$

- *Advantages*:
  - Highly adaptive to complex and dynamic operating conditions.
  - Potential for continuous improvement over time.
- *Disadvantages*:
  - Requires large datasets and computational resources.
  - Complexity in design and training.

The control of a bidirectional DC-DC converter interfaced with a supercapacitor is crucial for optimizing energy management in various applications, such as renewable energy systems and electric vehicles. This converter facilitates the efficient transfer of energy between the supercapacitor and the load or power source, allowing for both charging and discharging operations. Control strategies often employ advanced techniques such as proportional-integral (PI) control, model predictive control, or sliding mode control to regulate voltage and current levels, ensuring stability and responsiveness to dynamic load conditions. By effectively managing the supercapacitor's state of charge (SoC) and maintaining operational efficiency, these control systems enhance the overall performance, lifespan, and reliability of energy storage solutions while mitigating issues related to voltage fluctuations and energy losses. **Figure 12** presents a bidirectional control strategy for a DC-DC converter that is interfaced with a supercapacitor. This configuration is crucial for applications that require rapid charging and discharging capabilities, as supercapacitors can provide quick bursts of power to meet sudden energy demands or absorb excess energy during regenerative braking. The bidirectional control strategy allows for efficient management of energy flow, enabling the system to either draw energy from the supercapacitor to support immediate power needs or to charge the supercapacitor when excess energy is available. This flexibility enhances the overall performance and responsiveness of the energy storage system.

In contrast, **Figure 13** illustrates a similar bidirectional control strategy for a DC-DC converter interfaced with a battery. While batteries typically offer higher energy density and longer discharge durations compared to supercapacitors, they have slower response times. The bidirectional control in this context enables the battery to either supply power to the system during peak demand or be charged during low-demand periods or when excess energy is generated. This interaction helps maintain battery health and longevity by managing charge cycles effectively.

Together, these figures underscore the importance of tailored control strategies for different energy storage technologies within hybrid systems. By optimizing the interaction between supercapacitors and batteries through bidirectional converters, the overall efficiency, reliability, and performance of electric transportation systems can be significantly enhanced.

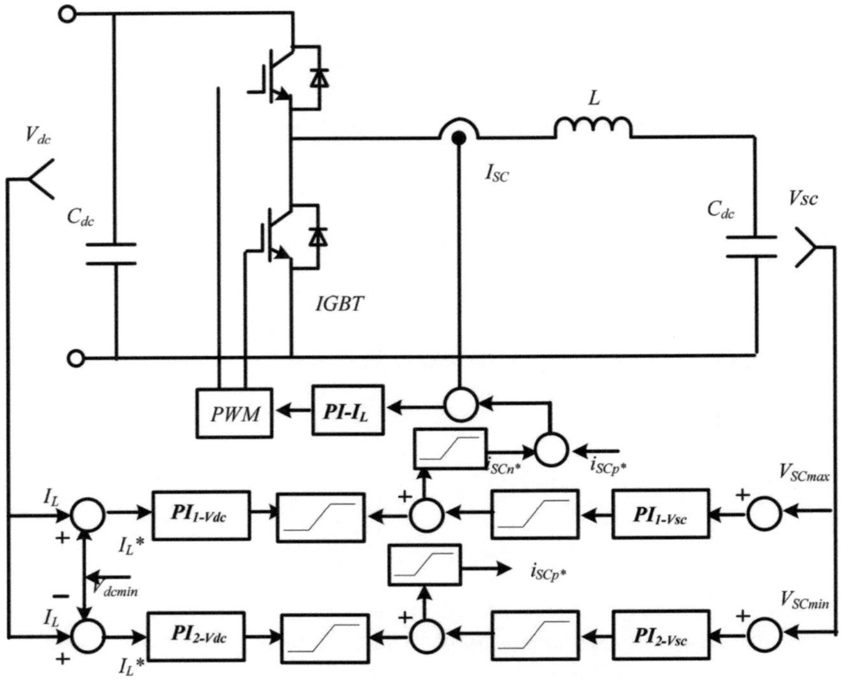

Figure 12.
Bidirectional control strategy for a DC-DC converter interfaced with a supercapacitor.

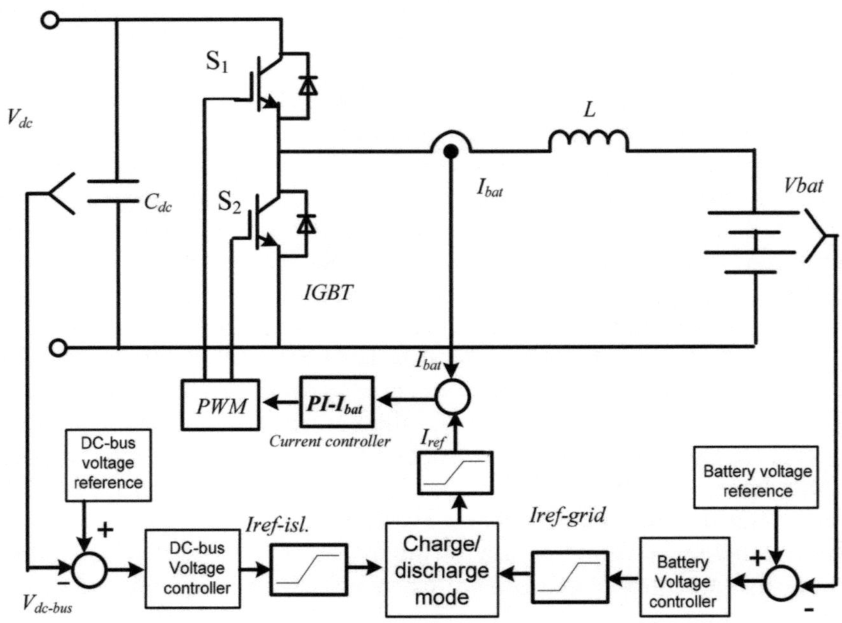

Figure 13.
Bidirectional control strategy for a DC-DC converter interfaced with a battery.

Scheme	Advantages	Disadvantages	Applications
Rule-based control	Simple and reliable	Limited adaptability	Basic HESS applications
Frequency decoupling	Effective in transient handling	Requires precise filter design	Systems with rapid load changes
Model predictive control	Optimized and adaptable	Computationally intensive	Advanced EVs and renewable systems
Fuzzy logic control	Handles nonlinearity	Relies on rule quality	Systems with uncertain dynamics
Power-weighted control	Real-time and straightforward	Requires accurate weights	General HESS applications
Droop control	Decentralized and scalable	Performance depends on tuning	Microgrids and distributed systems
AI-based control	Highly adaptive	Complex and resource-intensive	Cutting-edge EV and grid systems

Table 2.
Comparison of control schemes.

Table 2 provides a comprehensive comparison of various control schemes utilized in energy management systems, highlighting their advantages, disadvantages, and typical applications. Rule-based control is noted for its simplicity and reliability, making it suitable for basic hybrid energy storage system (HESS) applications, although it suffers from limited adaptability. Frequency decoupling excels in handling transients but requires precise filter design, making it ideal for systems experiencing rapid load changes. Model predictive control stands out for its optimization and adaptability; however, it is computationally intensive, which limits its application to advanced electric vehicles (EVs) and renewable energy systems. Fuzzy logic control effectively manages nonlinearity but relies heavily on the quality of the rules established, making it applicable in systems with uncertain dynamics. Power-WEIGHTED CONTROL offers real-time responsiveness and straightforward implementation but necessitates accurate weight assignments, fitting general HESS applications. Droop control is decentralized and scalable, yet its performance is heavily dependent on proper tuning, making it suitable for microgrids and distributed systems. Finally, AI-based control is highly adaptive and capable of handling complex scenarios, but it is also resource-intensive and complex, positioning it for cutting-edge EV and grid systems. This comparison underscores the diverse strengths and limitations of each control scheme, guiding practitioners in selecting the most appropriate solution for their specific energy management needs.

The choice of a control scheme depends on the system's complexity, performance requirements, and computational resources. For electric vehicles, schemes like frequency decoupling and model predictive control are popular due to their ability to handle dynamic loads efficiently.

Figure 14 illustrates a schematic diagram of microgrid control utilizing various energy storage systems (ESS). The diagram highlights the integration of multiple energy sources, including renewable resources such as solar panels and wind turbines, alongside conventional generators. It showcases how different ESS technologies, such as batteries, flywheels, and supercapacitors, are strategically deployed to enhance grid stability, optimize energy usage, and provide backup power during outages.

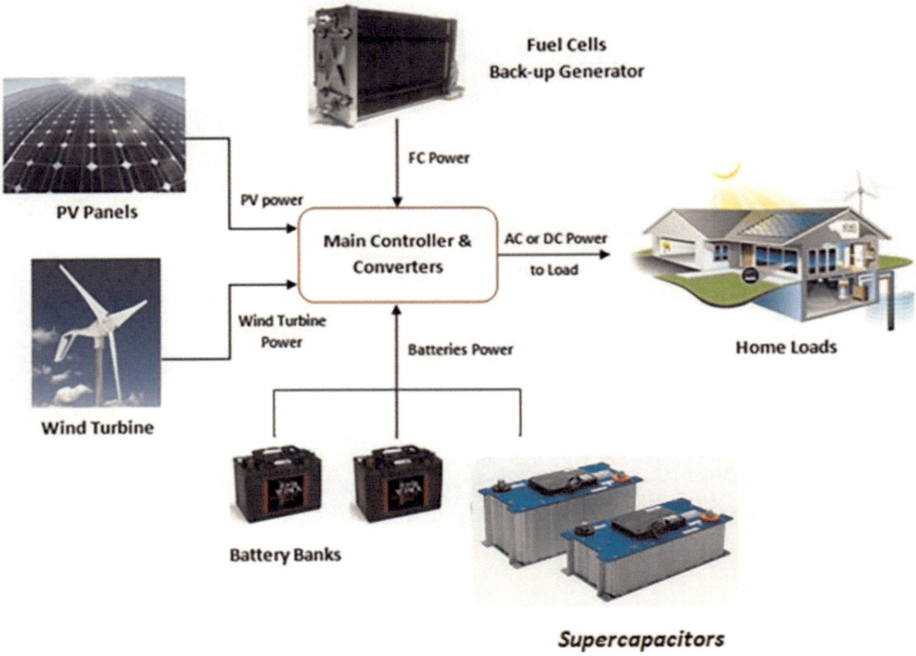

Figure 14.
Schematic diagram of microgrid control using different ESS.

The control architecture depicted emphasizes the role of advanced algorithms in managing the flow of energy, ensuring efficient load balancing, and facilitating seamless transitions between grid-connected and islanded modes of operation. Overall, this schematic serves as a comprehensive representation of the complexities involved in microgrid management and the importance of diverse energy storage solutions in achieving a resilient and sustainable energy system.

6.8 Advantages and challenges

Advantages:

- *High energy and power density*: Combines long-duration energy storage with rapid power delivery.
- *Extended battery life*: Reduces stress on batteries during high-power transients.
- *Fast charging*: Supercapacitors facilitate quick energy uptake and discharge.

Challenges:

- *Material costs*: Advanced hybrid systems may require expensive materials (e.g., graphene and carbon nanotubes).
- *System integration*: Effective control algorithms are needed to manage energy flow.

DOI: http://dx.doi.org/10.5772/intechopen.1010067

7. Energy management strategies

Hybrid energy storage systems (HESS) combine multiple energy storage technologies, such as batteries and supercapacitors, to leverage their complementary characteristics. Effective energy management strategies are essential for optimizing the interaction between these components, maximizing efficiency, extending lifespan, and ensuring optimal performance under varying load conditions. The strategies employed can vary based on the configuration of microgrids and specific applications, such as electric vehicles (EVs).

7.1 Energy management strategies in microgrid including DERs

7.1.1 Load prediction and demand response

One of the foundational strategies in HESS is load prediction, which involves forecasting the energy demand based on historical data and real-time monitoring. This allows the system to anticipate peak loads and adjust the energy distribution between the battery and supercapacitor accordingly. Demand response mechanisms can also be integrated to manage energy consumption dynamically, shifting usage patterns to reduce stress on the storage components.

7.1.2 State of charge (SoC) management

Effective SoC management is crucial for maintaining the health and longevity of both batteries and supercapacitors. Strategies involve continuously monitoring the SoC of each storage component and ensuring they operate within optimal ranges. For instance, supercapacitors can be used to handle short bursts of high power demand, while batteries can be reserved for longer-term energy supply. This approach minimizes deep cycling of the battery, extending its lifespan.

7.1.3 Power distribution control

Power distribution control strategies determine how much power is drawn from or supplied to each storage component based on real-time conditions. Techniques such as proportional control, where power is distributed according to the available SoC and power requirements, or more complex algorithms like model predictive control (MPC), which optimize future performance based on predicted load profiles, can be employed. These strategies ensure that the system responds efficiently to varying load conditions while balancing the charge and discharge rates of both components.

7.1.4 Energy routing algorithms

Energy routing algorithms help decide whether to draw power from the battery or supercapacitor based on efficiency and performance criteria. For example, when rapid bursts of power are needed, the system may prioritize supercapacitors due to their high discharge rates. Conversely, for sustained energy supply, the system would rely on batteries. Advanced algorithms can take into account factors such as efficiency losses, degradation rates, and operational costs to optimize energy routing.

7.1.5 Configuration-specific strategies

The configuration of microgrids plays a significant role in determining the appropriate energy management strategies. In decentralized microgrids, local energy sources (like solar or wind) can be integrated with HESS for better efficiency. Strategies may include:

- Decentralized control: Each storage unit operates independently based on local conditions.

- Centralized control: A central controller optimizes the overall system performance by coordinating between multiple storage units and generation sources.

- Hierarchical control: A combination of both decentralized and centralized approaches to balance local autonomy with global optimization.

Figure 15 illustrates a typical energy management system (EMS) structure for a standalone photovoltaic (PV) direct current (DC) microgrid that incorporates a parallel active hybrid energy storage system (HESS). In this configuration, the EMS plays a crucial role in coordinating the energy flows between the PV generation, the HESS, and the load demands. The parallel active HESS consists of multiple energy storage components, such as batteries and supercapacitors, which work in conjunction to optimize energy utilization. The EMS continuously monitors the state of charge (SoC) of each storage element and the power output from the PV system, dynamically adjusting the energy distribution to ensure efficient operation. This setup enhances the reliability and stability of the microgrid, allowing it to meet varying load demands while maximizing the use of renewable energy sources. By managing the interaction between generation, storage, and consumption effectively, the EMS contributes to improved energy efficiency and system resilience in standalone PV DC microgrids.

Figure 16 presents a comprehensive strategy for current determination and DC-bus voltage control within a microgrid system. This strategy is essential for ensuring stable operation and efficient power management. The current determination process involves monitoring real-time load demands and the available power from renewable sources, such as solar panels or wind turbines. By analyzing these parameters, the system can dynamically adjust the current supplied to various loads while maintaining the desired DC-bus voltage level. The DC-bus voltage control mechanism plays a pivotal role in this strategy, as it ensures that the voltage remains within specified limits, preventing potential damage to connected equipment and optimizing overall system performance. This dual approach not only enhances the reliability of the microgrid but also promotes efficient energy usage, facilitating a seamless integration of renewable energy sources into the power supply framework. Through effective current determination and voltage control, the microgrid can adapt to fluctuating energy demands and generation capacities, ultimately contributing to a more resilient and sustainable energy system.

7.2 Energy management strategies in electric vehicle (EV) powertrains

In EVs, energy management strategies are tailored to optimize the performance of the powertrain, which typically includes a combination of batteries and supercapacitors or other energy storage systems.

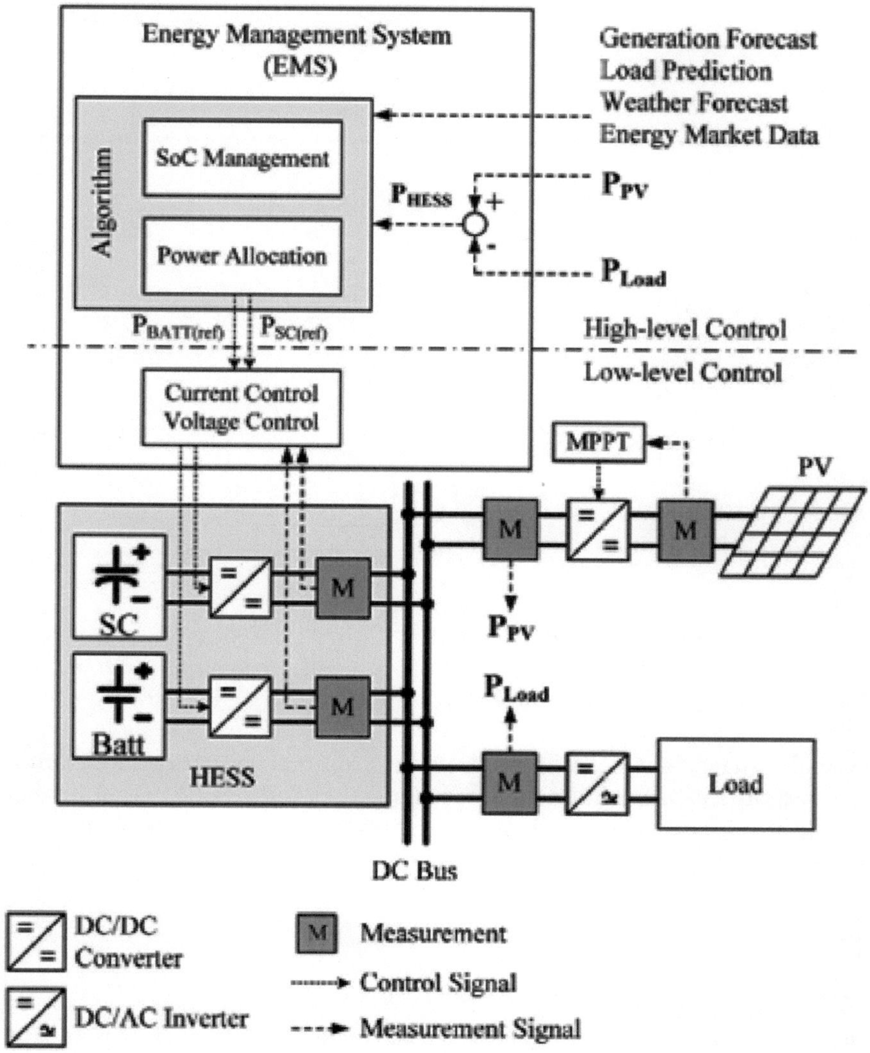

Figure 15.
Typical EMS structure for standalone PV DC microgrid with parallel active HESS.

7.2.1 Regenerative braking optimization

One critical strategy in EVs is optimizing regenerative braking, where kinetic energy from braking is converted back into electrical energy and stored in the battery or supercapacitor. Effective management ensures that this process maximizes energy recovery while preventing overcharging of the battery.

7.2.2 Dynamic power allocation

Dynamic power allocation strategies ensure that power demands are met efficiently during the acceleration, cruising, and deceleration phases. The system can switch between battery and supercapacitor based on current power needs—using the battery for sustained power and the supercapacitor for quick bursts.

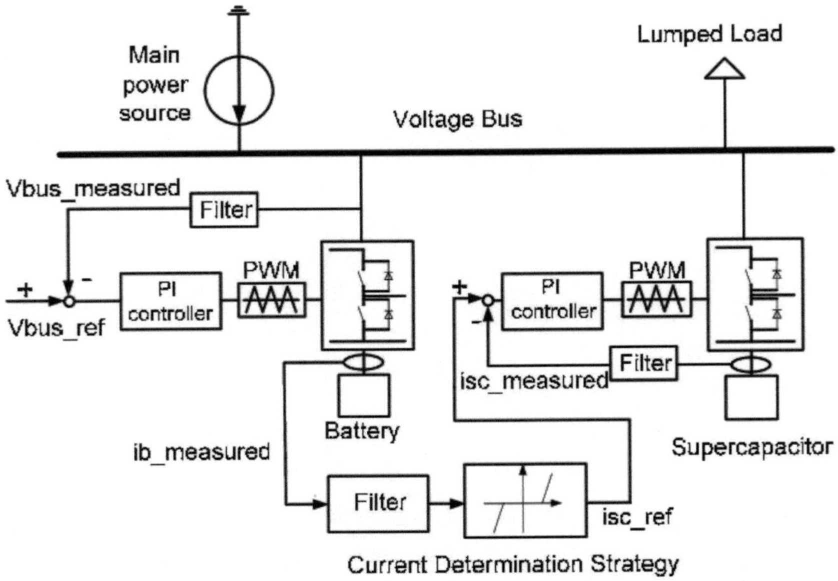

Figure 16.
Current determination strategy and DC-bus voltage control.

7.2.3 Thermal management integration

Thermal management is vital for maintaining optimal operating temperatures for batteries and supercapacitors. Energy management systems can incorporate temperature data to adjust charging and discharging rates, ensuring safety while maximizing performance.

7.2.4 Battery health monitoring

In EVs, continuous monitoring of battery health is essential for longevity. Energy management strategies include algorithms that predict battery degradation based on usage patterns and environmental conditions, allowing for proactive adjustments in charge cycles to extend battery life.

7.2.5 User-centric strategies

Incorporating user preferences into energy management can enhance user satisfaction. Strategies may involve learning driver habits to optimize energy use based on expected routes or integrating charging schedules that align with user availability.

Figure 17 illustrates a model of a hybrid energy storage system (HESS) designed to enhance energy management strategies for electric vehicles through the synergistic integration of batteries and supercapacitors. This innovative model focuses on developing a new hybrid electrochemical device known as the hybrid lithium-ion capacitor (HyLIC), which combines the strengths of lithium-ion batteries and lithium-ion capacitors. The HyLIC offers high energy density, prolonged cycle life, and exceptional power density, making it particularly suitable for the dynamic energy demands of electric vehicles. By improving thermal behavior in battery systems, this HESS model aims to optimize performance and efficiency, ultimately contributing to the advancement of electric vehicle technology and sustainable transportation solutions.

The design of the hybrid energy storage system, as depicted in **Figure 18**, showcases an innovative approach that integrates supercapacitors (SCs) and lithium-ion capacitors (LiCs), collectively referred to as hybrid capacitors (HCs), alongside a battery, utilizing a multiple input converter tailored for electric vehicles. This configuration leverages the strengths of both energy storage technologies, enabling enhanced energy density and power delivery while ensuring efficient energy management. The integration is made possible through a multiple-input DC-DC converter, which effectively coordinates the power flow between the battery and the hybrid capacitors, optimizing the overall performance and responsiveness of the energy storage system in dynamic driving conditions. This design aims to improve energy efficiency and extend the operational capabilities of electric vehicles, paving the way for more sustainable transportation solutions.

Figure 19 illustrates the hierarchical structure of control systems in energy storage and power sources, distinguishing between high-level and low-level control mechanisms. The high-level control focuses on strategic decision-making, such as optimizing energy flow, managing system efficiency, and ensuring grid stability. This layer typically involves algorithms that analyze data from various sources to make informed decisions regarding energy dispatch and storage management. In contrast, the low-level control is responsible for the real-time operation of individual components within the system, such as inverters, batteries, and power converters. This layer ensures that the commands from the high-level control are executed accurately and promptly, maintaining the operational integrity of the energy storage system (ESS) and its interaction with power sources.

Figure 20 depicts a power-sharing system designed to manage the distribution of energy from an energy storage system (ESS) to electric vehicle (EV) loads. This system is crucial for optimizing the charging process of EVs while balancing the available power from the ESS. The power-sharing mechanism ensures that energy is

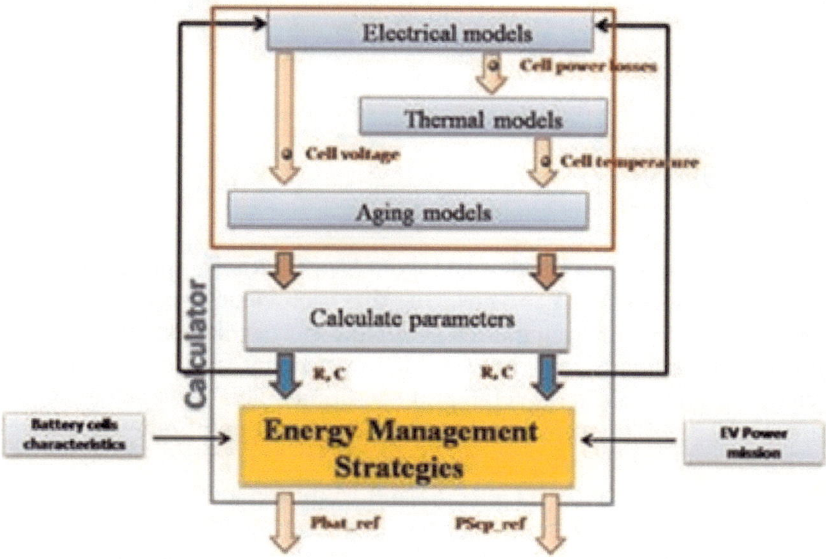

Figure 17.
A model of the hybrid energy storage system.

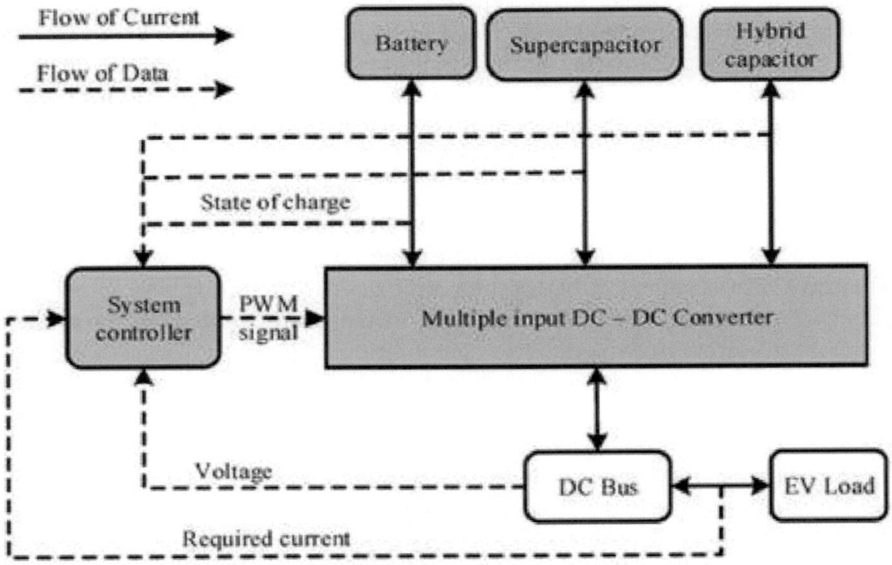

Figure 18.
Overview of the concept design.

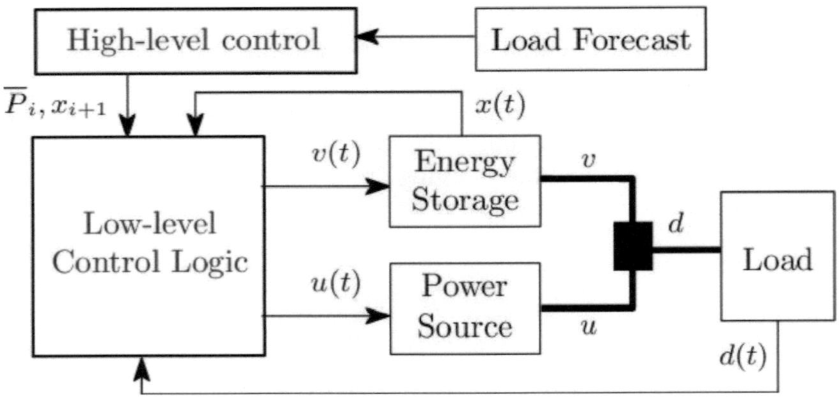

Figure 19.
High-level and low-level control of Energy storage and power source.

allocated efficiently based on the demand from EVs and the supply capabilities of the ESS, thus preventing overloads and ensuring that each vehicle receives adequate power for charging. By employing intelligent algorithms, this system can dynamically adjust power distribution in real-time, taking into account factors such as state of charge, charging priorities, and overall grid conditions. This approach enhances the overall efficiency of energy use and supports the integration of renewable energy sources with electric vehicle infrastructure.

Energy management strategies in HESS and EV powertrains are critical for optimizing performance, efficiency, and lifespan of energy storage components. By leveraging advanced algorithms and real-time data, these strategies ensure that systems can adapt to varying load conditions while minimizing wear on batteries and supercapacitors. The configuration of microgrids and specific application

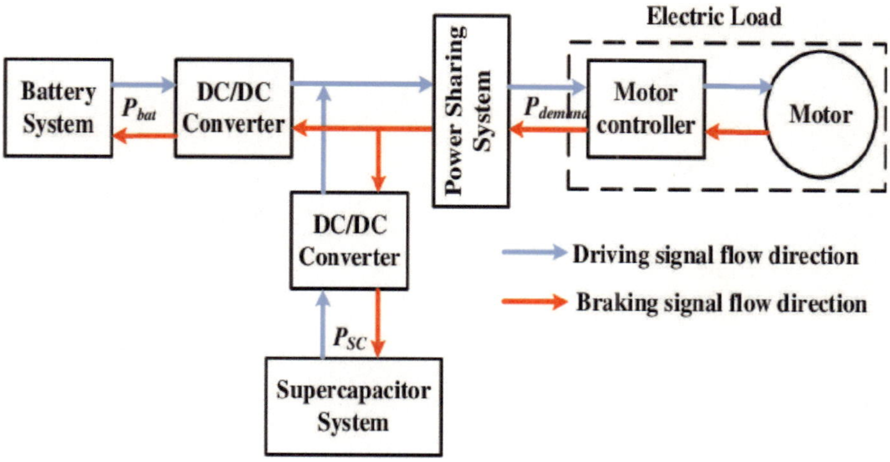

Figure 20.
Power sharing system of ESS to EV load.

requirements further tailor these strategies to meet diverse energy management needs effectively.

8. Performance evaluations of HESS

As mentioned, typical applications for each topology, showcase how different configurations can be more suitable for specific use cases—such as electric vehicles, renewable energy systems, or grid storage solutions. It would be beneficial if the figure included performance metrics or comparative analysis, illustrating how each topology fares in terms of energy density, power density, cycle life, and efficiency.

Figures 21 and **22** present a comparative analysis of bus voltage indices between battery energy storage systems (BESS) and hybrid energy storage systems (HESS). The graph highlights the performance differences in voltage stability and response under varying load conditions. BESS typically exhibits a more consistent bus voltage profile, reflecting its ability to provide steady power output, particularly during peak

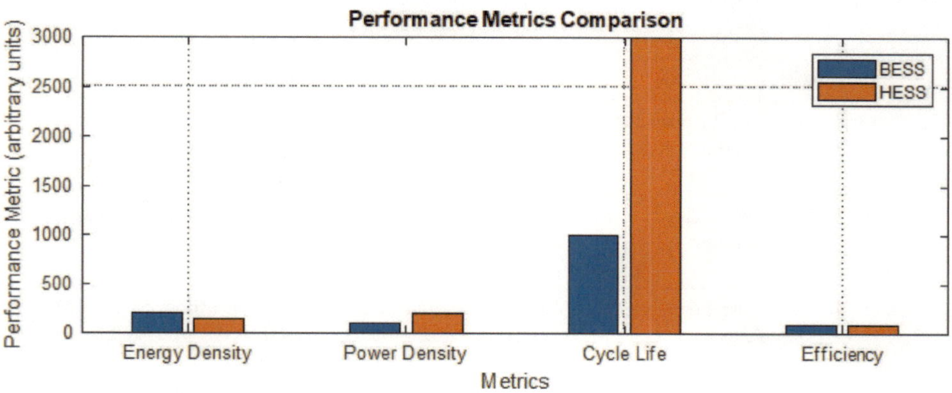

Figure 21.
Indices comparison between BESS and HESS.

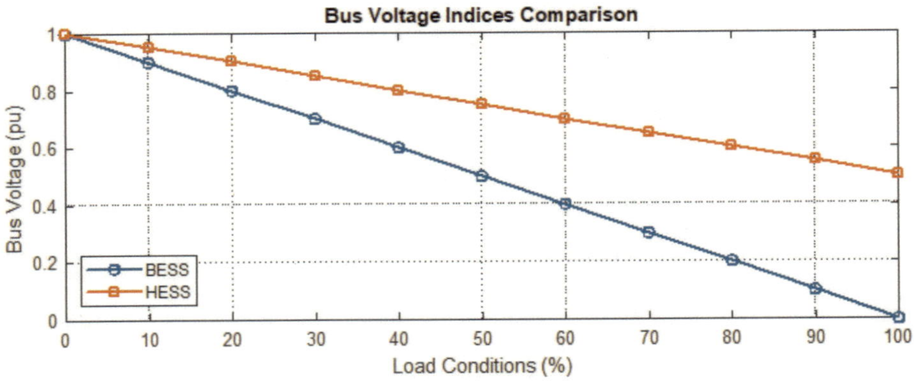

Figure 22.
Indices bus voltage comparison between BESS and HESS.

demand periods. In contrast, HESS demonstrates greater flexibility in voltage management due to its dual energy storage capabilities, which combine the rapid response of batteries with the longer-duration support of other energy storage technologies, such as supercapacitors or flywheels. This hybrid approach allows HESS to better accommodate fluctuations in load and generation, leading to improved overall voltage regulation. The indices presented in the figure underscore the advantages of integrating multiple storage technologies to enhance system reliability and performance in power distribution networks.

The generated figure provides a comparative analysis of the performance of battery energy storage systems (BESS) and hybrid energy storage systems (HESS) by evaluating bus voltage indices and key performance metrics, including energy density, power density, cycle life, and efficiency. The analysis is presented in two distinct visualizations: a bar chart and a line plot. The bar chart highlights the differences between BESS and HESS in terms of performance metrics, where example values are used for illustration and can be replaced with actual data for more precise insights. The line plot illustrates the bus voltage profiles under varying load conditions, demonstrating that while BESS maintains a consistent voltage profile, HESS exhibits dynamic adjustments owing to its hybrid energy storage design. These visualizations underscore the advantages of HESS in accommodating load fluctuations and enhancing overall system performance. **Table 3** shows the Quantitative Performance Comparison of hybrid topologies (passive, semi-active, active).

	Passive	Semi-active	Active
Efficiency	~85–90% efficiency due to lack of control over power flow.	~90–95% efficiency due to partial control.	95–98% efficiency due to optimized power flow management
Response time	Slow response due to fixed power sharing.	Faster response due to controlled supercapacitor discharge.	Fastest response due to dynamic power flow control
Cost	Lowest cost due to minimal components.	Moderate cost due to one power converter.	Highest cost due to multiple power converters and control systems

Table 3.
Quantitative Performance Comparison of hybrid topologies (passive, semi-active, active).

9. Integration of thermal and electrical energy storage

Thermal and electrical energy storage systems are being integrated to create *hybrid energy solutions* capable of addressing diverse energy requirements.

9.1 Concept and design

- *Thermal energy storage* (*TES*): Utilizes phase change materials (PCMs), molten salts, or sensible heat systems to store heat energy.
- *Electrical energy storage* (*EES*): Batteries and supercapacitors deliver electrical energy on demand.

Hybrid systems use *waste heat recovery* from electrical energy storage processes (e.g., batteries) to improve efficiency and reduce energy losses.

9.2 Applications

- *Renewable energy integration*: TES stores thermal energy from concentrated solar power (CSP), while batteries manage electrical loads.
- *Combined heat and power* (*CHP*): Hybrid systems deliver simultaneous thermal and electrical energy for industrial processes.

10. Future trends and outlook

The development of *hybrid and advanced energy storage systems* is accelerating due to the following trends:

1. *Advanced materials*: Continued innovation in MXenes, MOFs, and other nanomaterials.
2. *Energy management systems*: AI-based control algorithms for optimal hybrid system performance. AI and machine learning are already being used for predictive maintenance, fault detection, and real-time optimization of energy storage systems.
    - Use of reinforcement learning and deep learning to develop adaptive control algorithms that optimize power flow between storage components in real time.
    - Implementation of multi-objective optimization techniques to balance efficiency, cost, and lifespan.
    - AI-driven models to forecast energy demand, renewable energy generation, and storage system performance.
    - Integration of weather data and load profiles to improve system reliability and efficiency.

- Creation of digital replicas of HESS to simulate and optimize system behavior under various conditions.
- Use of digital twins for predictive maintenance and performance enhancement.
- Development of self-learning HESS that can autonomously adapt to changing operating conditions and user requirements.
- Integration of AI with IoT devices for seamless communication and control in smart grids and microgrids.

3. *Scalability*: Hybrid storage systems will play a critical role in stabilizing grids with high penetration of intermittent renewable energy sources (e.g., solar and wind).

- Increased deployment of HESS in microgrids and utility-scale projects to balance supply and demand.
- Development of advanced control algorithms to optimize energy storage and renewable energy integration.
- Use of predictive analytics and AI to forecast energy generation and consumption patterns.

4. *Sustainability*: Focus on recyclable and low-cost materials to reduce environmental impacts. There will be a stronger focus on sustainability and recycling in the design and deployment of hybrid storage systems.

- Development of eco-friendly materials and manufacturing processes for energy storage components.
- Implementation of closed-loop recycling systems to recover and reuse critical materials (e.g., lithium and cobalt).
- Increased use of second-life batteries in hybrid storage systems for stationary applications.

10.1 Materials for high-energy-density supercapacitors

The development of advanced materials is crucial for enhancing the performance of supercapacitors, particularly in achieving higher energy densities. MXenes, a family of two-dimensional transition metal carbides and nitrides, have emerged as a promising option due to their exceptional properties. With high electrical conductivity and a large surface area, MXenes facilitate efficient ion adsorption, significantly improving charge storage capabilities. Their stability in electrochemical applications makes them suitable for a variety of uses, including electric vehicles (EVs), grid storage solutions, and portable electronics. Similarly, Metal-Organic Frameworks (MOFs) present another innovative avenue for supercapacitor development. These highly porous materials can be engineered with tunable structures to optimize both energy and power densities. When combined with conductive materials such as graphene, hybrid

MOFs can further enhance charge storage, making them ideal candidates for next-generation supercapacitors.

10.2 Hybrid supercapacitors

Hybrid supercapacitors represent a significant advancement in energy storage technology by integrating faradaic (battery-like) and non-faradaic (capacitor-like) materials. This combination allows for higher energy density while maintaining the rapid charge-discharge rates characteristic of traditional supercapacitors. Key features of hybrid supercapacitors include the use of asymmetric electrodes that incorporate battery-type materials, such as lithium cobalt oxide, alongside supercapacitor electrodes like activated carbon. Additionally, advanced electrolytes, including ionic liquids and gel electrolytes, are employed to enhance operating voltage and overall system stability. The versatility of hybrid supercapacitors opens up numerous applications, particularly in wearable electronics where flexible and lightweight designs are essential for smart devices. They also play a critical role in electric vehicles and transportation systems, providing high-power assist capabilities for busses, trains, and drones. Moreover, their rapid response to fluctuations in renewable energy generation makes them invaluable for energy storage solutions that support solar and wind technologies.

11. Conclusion

Hybrid and advanced energy storage systems represent a transformative solution to the challenges of modern energy applications. Battery-supercapacitor hybrids, thermal-electric systems, and high-performance supercapacitors combine to deliver flexible, scalable, and efficient energy storage. These technologies play a pivotal role in electric vehicles, renewable energy integration, smart infrastructure, and IoT systems, offering a sustainable path toward the energy future.

Hybridization of supercapacitors with battery systems can overcome the limitations in both the thermodynamics and kinetics of the electrochemical reactions involved in the battery technologies as they do not fully meet the requirements of irregular energy consumption of vehicles. Improved supercapacitors and their variants present huge opportunities in minigrids, trains, trams, trucks, heavy off-road vehicles, tiny uninterruptable power supplies for IoT nodes and 1 MWh giants for hospitals and data centers. They already drive brain scanners, lifts, Maglev trains, power rail, laser guns, vehicle brakes, aircraft and bus doors. Currently researchers are working on the opportunities for designing native battery/supercapacitor systems and development of eco-friendly hybrid and electric vehicles. Challenges still exist in the form of requirement of novel electrode materials, such as silicon-based, nanocarbon-based, etc., implementation of novel coating methodologies such as 3D printing and inkjet printing to reduce the thickness, and so on; so as to promote utilization of supercapacitors on a commercial scale.

Author details

Khairy Sayed[1], Mahmoud Aref[2], Ahmed G. Abokhalil[1,3] and Mahmoud A. Mossa[4*]

1 Electrical Engineering Department, Faculty of Engineering, Sohag University, Sohag, Egypt

2 Department of Electrical Engineering, Assiut University, Assiut, Egypt

3 Department of Sustainable and Renewable Energy Engineering, College of Engineering, University of Sharjah, Sharjah, United Arab Emirates

4 Electrical Engineering Department, Faculty of Engineering, Minia University, Minia, Egypt

*Address all correspondence to: mahmoud_a_mossa@mu.edu.eg

References

[1] Khalid M. A review on the selected applications of battery-supercapacitor hybrid energy storage systems for microgrids. Energies. 2019;**12**(23):4559

[2] Hossein R, Seyed EA, Seyedreza A, Shaahin F. Energy management strategies of battery-ultracapacitor hybrid storage systems for electric vehicles: Review, challenges, and future trends, Journal of Energy Storage. 2022; **53**:105045. ISSN 2352-152X. DOI: 10.1016/j.est.2022.105045

[3] Sayed K et al. Feasibility study and economic analysis of PV/wind-powered hydrogen production plant. IEEE Access. 2024;**12**:76304-76318. DOI: 10.1109/ACCESS.2024.3406895

[4] Mesbah MA, Sayed K, Ahmed A, Lmalki MM, Alghamdi TAH. A distributed architecture of parallel buck-boost converters and cascaded control of DC microgrids-real time implementation. IEEE Access. 2024;**12**: 47483-47493

[5] Sayed K, El-Zohri HH, Ahmed A, Khamies M. Application of tilt integral derivative for efficient speed control and operation of BLDC motor drive for electric vehicles. Fractal and Fractional. 2024;**8**:61. DOI: 10.3390/fractalfract8010061

[6] Mesbah MA, Sayed K, Ahmed A, Aref M, Elbarbary ZMS, Almuflih AS, et al. Adaptive control approach for accurate current sharing and voltage regulation in DC microgrid applications. Energies. 2024;**17**:284. DOI: 10.3390/en17020284

[7] Ebeed M, Abdelmotaleb MA, Khan NH, Jamal R, Kamel S, Hussien AG, et al. A modified artificial hummingbird algorithm for solving optimal power flow problem in power systems. Energy Reports. 2024;**11**: 982-1005

[8] Elnozahy A, Mohamed M, Sayed K, Bahyeldin M, Mohamed SA. Fault identification and classification algorithm for high voltage transmission lines based on a fuzzy-neuro-fuzzy approach. International Journal of Modelling and Simulation. 2023. pp. 1-12. DOI: 10.1080/02286203.2023.2274062

[9] Ahmed NM, Ebeed M, Magdy G, Sayed K, Gamoura SC, Metwally ASM, et al. A new optimized FOPIDA-FOIDN controller for the frequency regulation of hybrid multi-area interconnected microgrids. Fractal and Fractional. 2023; **7**:666. DOI: 10.3390/fractalfract7090666

[10] Abo-Khalil AG, Sayed K, Radwan A, El-Sharkawy IIA. Analysis of the PV system sizing and economic feasibility study in a grid-connected PV system. Case Studies in Thermal Engineering. 2023;**45**:102903. DOI: 10.1016/j.csite.2023.102903

[11] Alrumayh O, Sayed K, Amutairi A. LVRT and reactive power/voltage support of utility-scale PV power plants during disturbance conditions. Energies. 2023;**16**(7):3245. DOI: 10.3390/en16073245

[12] Sayed K, Abdel-Khalek S, Zakaly HMH, Aref M. Energy management and control in multiple storage energy units (battery–supercapacitor) of fuel cell electric vehicles. Materials. 2022;**15**:8932. DOI: 10.3390/ma15248932

[13] Sayed K, Abo-Khalil AG. An Interleaved Two Switch Soft-Switching Forward PWM Power Converter with

Current Doubler Rectifier. Electronics. 2022;**11**(16):2551. DOI: 10.3390/electronics11162551

[14] Sayed K, Almutairi N, Albagami N, Alrumayh O, Abo-Khalil AG, Saleeb H. A review of DC-AC converters for electric vehicle applications. Energies. 2022;**15**: 1241. DOI: 10.3390/en15031241

[15] Saleeb H, Kassem R, Sayed K. Artificial neural networks applied on induction motor drive for an electric vehicle propulsion system. Electrical Engineering. 2022;**104**(3):1769-1780

[16] Melkia C, Ghoudelbourk S, Soufi Y, Maamri M, Bayoud M. Battery-Supercapacitor hybrid energy storage Systems for Stand-Alone Photovoltaic. European Journal of Electrical Engineering. **24**(5-6):265-271. DOI: 10.18280/ejee.245-605

[17] Xing F, Bi Z, Feng S, Liu F, Zhong-Shuai W. Unraveling the design principles of battery-supercapacitor hybrid devices: From fundamental mechanisms to microstructure engineering and challenging perspectives. Advanced Energy Materials. 2022;**12**(26):24. DOI: 10.1002/aenm.202200594

[18] Dam DH, Lee HH. Battery–inductor–supercapacitor hybrid energy storage system for DC microgrids. Journal of Power Electronics. 2020;**20**: 308-318. DOI: 10.1007/s43236-019-00027-0

[19] Shan W, Schwalm M, Shan M. A design tool for battery/supercapacitor hybrid energy storage systems based on the physical–electrochemical degradation battery model basis. Energies. 2024;**17**:3481. DOI: 10.3390/en17143481

[20] Hassan MR, Mossa MA, Dousoky GM. Evaluation of electric dynamic performance of an electric vehicle system using different control techniques. Electronics. 2021;**10**(21): 2586

[21] Abo-Khalil AG, Sayed K, Elnozahy A, Yu BG. Modeling and power optimization control of tidal energy systems. Journal of Engineering Research (Kuwait). 2022;**10**(1):190-202

[22] Almutairi A, Sayed K, Albagami N, Abo-Khalil AG, Saleeb H. Multi-port PWM DC-DC power converter for renewable energy applications. Energies. 2021;**14**(12):3490. DOI: 10.3390/en14123490

[23] AbouKhalil AG, Eltamaly AM, Alsaud MS, Sayed K, Alghamdi AS. Sensorless control for PMSM using model reference adaptive system. International Transactions on Electrical Energy Systems. 2021;**31**(2):e12733. DOI: 10.1002/2050-7038.12733

[24] Sayed K, Kassem A, Saleeb H, Alghamdi AS, Abo-Khalil AG. Energy-saving of battery electric vehicle powertrain and efficiency improvement during different standard driving cycles. Sustainability. 2020;**12**:10466

[25] Almutairi A, Abo-Khalil AG, Sayed K, Albagami N. MPPT for a PV grid-connected system to improve efficiency under partial shading conditions. Sustainability. 2020;**12**: 10310

[26] Alghamdi AS, Sayed K, Abokhalil AG, Awan AB, Zohdy MA. A soft switching DC-link quasi resonant three-phase inverter for AC servo-motor drive applications. International Transaction Journal of Engineering, Management, & Applied Sciences & Technologies. 2020;**12**(1):12A1N, 1-13. Available from: http://TUENGR.COM/

V12/12A1N.pdf. DOI: 10.14456/ITJEMAST.2021.14

[27] Eltamaly AM, Al-Saud M, Sayed K, Abo-Khalil AG. Sensorless active and reactive control for DFIG wind turbines using opposition-based learning technique. Sustainability. 2020;**12**:3583. DOI: 10.3390/su12093583

[28] Kassem R, Sayed K, Kassem A, Mostafa R. Power optimisation scheme of induction motor using FLC for electric vehicle. IET Electrical Systems in Transportation;**10**(3):301-309. DOI: 10.1049/iet-est.2019.0151

[29] Praveen RP, Therattil J, Jose J, Abo-Khalil A, Alghamdi A, Bindu GR, et al. Hybrid control of a multi area multi machine power system with FACTS devices using non-linear modelling. IET Generation, Transmission & Distribution. 2020;**14**(10):1993-2003

[30] Sayed K, Ali ZM, Aldhaifallah M. Phase-shift PWM-controlled DC–DC converter with secondary-side current doubler rectifier for on-board charger application. Energies. 2020;**13**:2298

[31] Sayed K, Gronfula MG, Ziedan HA. Novel soft-switching integrated boost DC-DC converter for PV power system. Energies. 2020;**13**(3):749, 1-17

[32] Ashraf N, Izhar T, Abbas G, Awan AB, Alghamdi AS, Abo-Khalil AG, et al. A new single-phase direct frequency controller having reduced switching count without zero-crossing detector for induction heating system. Electronics. 2020;**9**:430. DOI: 10.3390/electronics9030430

[33] Saleeb H, Sayed K, Kassem A. et al. Control and analysis of bidirectional interleaved hybrid converter with coupled inductors for electric vehicle applications. Electrical Engineering. 2020;**102**:195-222. DOI: 10.1007/s00202-019-00860-3

[34] Sayed K, Abo-Khalil AG, Alghamdi AS. Optimum resilient operation and control DC microgrid based electric vehicles charging station powered by renewable energy sources. Energies. 2019;**12**:4240. DOI: 10.3390/en12224240

[35] Sayed K. ZVS soft-switched DC-DC converter based charger for low voltage battery in hybrid electric vehicles. IET Power Electronics. 2019;**12**:3389-3396

[36] Iskander Tlili M, Osman EM, Barhoumi IA, Abo-Khalil AG, Praveen RP, Sayed K. Performance enhancement of a humidification–dehumidification desalination system. Journal of Thermal Analysis and Calorimetry. 2020;**140**:309-319

[37] Abo-Khalil AG, Alghamdi AS, Eltamaly AM, Al-Saud MS, Praveen RP, Sayed K. Design of state feedback current controller for fast synchronization of DFIG in wind power generation systems. Energies. 2019;**12**(12):2427

[38] Abo-Khalil AG, Alyami S, Sayed K, Alhejji A. Dynamic modeling of wind turbines based on estimated wind speed under turbulent conditions. Energies. 1907;**12**(10):2019

[39] Saleeb H, Sayed K, Kassem A, Mostafa R. Power management strategy for battery electric vehicles. IET Electrical Systems in Transportation. 2019;**9**(2):65-74

[40] Abdel-Salam M, Kamel R, Sayed K, Khalaf M. Design and implementation of a multifunction DSP-based-numerical relay. Electric Power Systems Research. 2017;**143**:32-43

[41] Sayed K, Abdel-Salam M. Dynamic performance of wind turbine conversion system using PMSG-based wind simulator. Electrical Engineering Journal. 2017;**99**:431-439

[42] Sayed K. A high efficiency DC-DC converter with LC resonant in the load-side of HFT and voltage doubler for solar PV systems. International Journal of Power Electronics. 2017;**8**(3):232-248

[43] Sayed K, Mahfouz H, Elzohry E. Analysis and design for interleaved ZCS buck DC-DC converter with low switching losses. International Journal of Power Electronics. 2017;**8**(3):210-231

[44] Sayed K, Gabbar HA. Supervisory control of a resilient DC microgrid for commercial buildings. International Journal of Process Systems Engineering. 2017;**4**(2–3):99-118

[45] Abdel-Salam M, Sayed K, Ahmed A, Amery M, Swify M. Design, implementation and operation of a standalone residential photovoltaic system. International Journal of Power and Energy Conversion. 2017;**8**(1):47-67

[46] Sayed K, Gabbar HA. Electric vehicle to power grid integration using three-phase three-level AC/DC converter and PI-fuzzy controller. Energies. 2016;**9**(7):532

[47] Mesbahi T, Rizoug N, Bartholomeüs P, Le Moigne P. A new energy management strategy of a Battery/ Supercapacitor Hybrid Energy Storage System for electric vehicular applications. In: 7th IET International Conference on Power Electronics, Machines and Drives (PEMD 2014). Manchester, UK; 2014. pp. 1-7. DOI: 10.1049/cp.2014.0442

[48] Hannan MA, Hoque MM, Mohamed A, Ayob A. Review of energy storage systems for electric vehicle applications: Issues and challenges. Renewable and Sustainable Energy Reviews. 2017;**69**:771-789

[49] Kouchachvili L, Yaïci W, Entchev E. Hybrid battery/supercapacitor energy storage system for the electric vehicles. Journal of Power Sources. 2018;**374**: 237-248

[50] Rizoug N, Mesbahi T, Sadoun R, Bartholomeüs P, Le Moigne P. Development of new improved energy management strategies for electric vehicle battery/supercapacitor hybrid energy storage system. Energy Efficiency. 2018; **11**(4):823-843

[51] CVV MG, Ramesh R. Review of battery-supercapacitor hybrid energy storage systems for electric vehicles. Results in Engineering. 2024;**24**:103598. ISSN 2590-1230. DOI: 10.1016/j.rineng.2024.103598

[52] Pant YV, Nghiem TX, Mangharam R. Peak Power Control of Battery and Super-capacitor Energy Systems in Electric Vehicles. Available from: https://www.researchgate.net/publication/304088308

[53] Song Z, Li J, Hou J, Hofmann H, Ouyang M, Jiuyu D. The battery-supercapacitor hybrid energy storage system in electric vehicle applications: A case study. Energy. 2018;**154**:433-441. DOI: 10.1016/j.energy.2018.04.148

[54] Jing W, Lai CH, Wong SHW, Wong MLD. Battery-supercapacitor hybrid energy storage system in standalone DC microgrids: A review. IET Renewable Power Generation. 2017; **11**(4):461-469. DOI: 10.1049/iet-rpg.2016.0500

[55] Montenegro-Oviedo JA, Ramos-Paja CA, Orozco-Gutierrez ML, Franco-

Mejía E, Serna-Garcés SI. Design and experimental validation of a battery/supercapacitor hybrid energy storage system based on an adaptive LQG controller. Applied System Innovation. 2025;**8**:1. DOI: 10.3390/asi8010001

[56] Zhang P, Zheng Y, Wang H, Jin-Ming W, Zhang Z, Wen W. A battery-supercapacitor hybrid energy storage device that directly uses seawater or saltwater lake water. Materials Today Advances. 2024;**24**:100535. ISSN 2590-0498. DOI: 10.1016/j.mtadv.2024.100535

[57] Paul T, Mesbahi T, Durand S, Flieller D, Uhring W. Sizing of lithium-ion battery/supercapacitor hybrid energy storage system for forklift vehicle. Energies. 2020;**13**:4518. DOI: 10.3390/en13174518

[58] Iqbal MZ, Aziz U. Supercapattery: Merging of battery-supercapacitor electrodes for hybrid energy storage devices. Journal of Energy Storage. 2022;**46**:103823. DOI: 10.1016/j.est.2021.103823

[59] Baboo JP, Jakubczyk E, Yatoo MA, Phillips M, Grabe S, Dent M, et al. Investigating battery-supercapacitor material hybrid configurations in energy storage device cycling at 0.1 to 10C rate. Journal of Power Sources. 2023;**561**:232762. DOI: 10.1016/j.jpowsour.2023.232762

[60] Reddy RM, Das M, Chauhan N. Novel battery-supercapacitor hybrid energy storage system for wide ambient temperature electric vehicles operation. IEEE Transactions on Circuits and Systems II: Express Briefs. 2023;**70**(7):2580-2584. DOI: 10.1109/TCSII.2023.3237860

[61] Santhi Maria Benoy, Mayank Pandey, Dhrubajyoti Bhattacharjya, Binoy K. Saikia, Recent trends in supercapacitor-battery hybrid energy storage devices based on carbon materials. Journal of Energy Storage. 2022;**52**(Part B):104938. ISSN 2352-152X. DOI: 10.1016/j.est.2022.104938

[62] Zhang Y, Jiang Z, Yu X. Control Strategies for Battery/Supercapacitor Hybrid Energy Storage Systems. In: 2008 IEEE Energy 2030 Conference. Atlanta, GA, USA; 2008. pp. 1-6. DOI: 10.1109/ENERGY.2008.4781031

[63] Kouchachvili L, Yaïci W, Entchev E. Hybrid battery/supercapacitor energy storage system for the electric vehicles. Journal of Power Sources. 2018;**374**:237-248. DOI: 10.1016/j.jpowsour.2017.11.040

[64] Muralee CVV, Gopi RR. Review of battery-supercapacitor hybrid energy storage systems for electric vehicles. Results in Engineering. 2024;**24**:103598. ISSN 2590-1230. DOI: 10.1016/j.rineng.2024.103598

[65] Abro GEM, Zulkifli SAB, Kumar K, El Ouanjli N, Asirvadam VS, Mossa MA. Comprehensive review of recent advancements in battery technology, propulsion, power interfaces, and vehicle network systems for intelligent autonomous and connected electric vehicles. Energies. 2023;**16**(6):2925, p. 4

[66] Liu F, Wang C, Luo Y. Parameter matching method of a battery-supercapacitor hybrid energy storage system for electric vehicles. World Electric Vehicle Journal. 2021;**12**:253. DOI: 10.3390/wevj12040253

[67] Hassan MRM, Mossa M, Dousoky GM. Dynamic performance analysis of an electric vehicle system using different control algorithms.

Journal of Advanced Engineering Trends. 2024;**43**(1):151-162. DOI: 10.21608/jaet.2022.138420.1202

[68] Ren Y, Rind SJ, Jiang L. A coordinated control strategy for battery/supercapacitor hybrid energy storage system to eliminate unbalanced voltage in a standalone AC microgrid. Journal of Intelligent Manufacturing and Special Equipment. 2020;**1**(1):3-23. ISSN: 2633-6596

[69] Mossa MA, Gam O, Bianchi N, Quynh NV. Enhanced control and power management for a renewable energy-based water pumping system. IEEE Access. 2022;**10**:36028-36056. DOI: 10.1109/ACCESS.2022.3163530

[70] Ebeed M et al. Performance enhancement of gulf of El Zayt wind farm in Egypt using a novel optimized seconded sliding mode control. IEEE Access. 2024;**12**:185605-185617. DOI: 10.1109/ACCESS.2024.3510252

[71] Mossa MA, Gam O, Bianchi N. Dynamic performance enhancement of a renewable energy system for grid connection and stand-alone operation with battery storage. Energies. 2022; **15**(3):1002. DOI: 10.3390/en15031002

[72] Mossa MA, Echeikh H, Quynh NV, Bianchi N. Performance dynamics improvement of a hybridwind/fuel cell/battery system for standalone operation. IET Renewable Power Generation. 2022; **17**(2):349-375. DOI: 10.1049/rpg2.12603

[73] Chojaa H et al. Robust control of DFIG-based WECS integrating an energy storage system with intelligent MPPT under a real wind profile. IEEE Access. 2023;**11**:90065-90083. DOI: 10.1109/ACCESS.2023.3306722

Chapter 2

Optimizing Energy Storage Solutions for Grid Resilience: A Comprehensive Overview

Sina Samadi Gharehveran, Kimia Shirini and Arya Abdolahi

Abstract

The evolving energy landscape, driven by increasing demands and the growing integration of renewables, necessitates a dynamic adjustment of the energy grid. To enhance the grid's resilience and accommodate the surging influx of green energy. Energy storage solutions have emerged as crucial components. Despite considerable research, there remains a notable gap in systematically assessing the suitability of different storage devices across diverse stationary applications. This review endeavors to bridge this gap by thoroughly examining the current landscape of energy storage and discerning its aptness for various grid support applications. Through an exploration of technical, economic, and environmental considerations, the study aims to elucidate the optimal storage technologies for different contexts. Among electrochemical storage options, lithium-ion batteries emerge as optimal choices for both low- and medium-scale applications, owing to their robust power and energy densities. Meanwhile, capacitors, supercapacitors, and superconductive magnetic energy storages exhibit promise for high-power demands within the electrical storage domain. Additionally, thermal energy storage presents a viable solution for seasonal and bulk energy requirements. This review suggests using a mix of technologies in hybrid solutions to better meet the unique needs of different applications.

Keywords: energy storage, renewable integration, techno-economic assessment, hybrid solutions, green energy

1. Introduction

The world continues to grapple with significant energy challenges, including the scarcity of reliable energy sources at reasonable costs and the environmental damage caused by polluting sources like coal. To address these issues, many countries are increasingly adopting various forms of renewable energy sources (RESs). It is projected that by 2050, wind and solar energy will fulfill 50% of global energy demand [1]. Additionally, the demand for electricity from electric vehicles (EVs) is expected to grow by 6%, reaching approximately 2 TWh by 2040 [2]. Based on the Bloomberg New Energy Finance (BNEF) report examining the global power generation mix, fossil fuels dominated the energy supply from 1970 to 2017, significantly outweighing renewable sources. However, since 2018, the reliance on fossil fuels such as coal and natural gas has been decreasing, with projections indicating a decline to 31% of the total energy mix by 2050. Conversely, renewable energy sources such as hydro, wind, and solar are expected to dominate, comprising over 62% of the energy mix. Specifically, solar and wind are anticipated to account for around 48% of the generation mix. This transition is propelling a significant surge in the deployment of network support and energy storage solutions globally. A recent BNEF study projects that by 2040, the market will grow to 1095 GW/2850 GWh, representing over a 120-fold increase from the 9 GW/17 GWh installed in 2018 [3].

Energy storage devices (ESDs) can be utilized across all levels of the network, including generation, transmission, distribution, and for local industrial and commercial customers. Presently, there is a growing interest not only in deploying existing ESDs for stationary applications but also in developing advanced future ESDs. This category encompasses upcoming advancements such as Li-ion, solid-state, lithium-polymer, lithium-sulfur, lithium-metal-polymer, metal-ion batteries, organic radical batteries, hybrid supercapacitors, and other emerging technologies [4, 5]. Studies [6, 7] indicate that for large-scale power management, thermal energy storage (TES) is currently a viable option, particularly for capacities exceeding several MW. Among electric and electrochemical ESDs, only flow batteries, sodium-sulfur, and lead-acid batteries are considered capable of meeting these extensive requirements. Additionally, various ESDs are suitable for integrating green energy at the distribution and transmission levels, addressing diverse multidisciplinary needs [6, 7]. Assessing the present and prospective states of electrical, electrochemical, and thermal ESDs is essential for determining their suitability for various roles and

		Integrated centralized storage and transmission framework	**The distribution system and regional storage**	**Consumer (building and residential level)**
Function	Balance between supply and demand	Large geographic disparities and variations on seasonal, weekly, daily, and hourly scales result from the intermittent nature of wind turbines (WT) and solar photovoltaics (PV), leading to fluctuating electricity generation	Daily/hourly variations Peak saving	Daily deviations
	Distribution—moving energy	Voltage and frequency regulation Increased peak generation for traditional power stations Global electricity market	Voltage and frequency control Power market	Combining small stored energy quantities to address distribution needs, including capacity issues and reducing losses
	Energy efficiency improvement	Improved productivity in the globe energy mix with time-shift	Storage and load control for improved performance in the distribution system	Enhanced energy production and utilization, along with changes in behaviors

Table 1.
Function of energy storage in the energy supply chain [8].

power applications. **Table 1** outlines the roles and significance of ESDs, their required specifications, and how they interact across various power levels.

This review distinguishes itself from prior reports [4, 7, 9] in several key aspects. It provides a comprehensive analysis of ESD using up-to-date and reliable data. In contrast to earlier reviews, this study integrates graphical analysis and compares it with effective and standardized performance evaluation criteria, encompassing economic, technical, and environmental metrics. In terms of economics, this review goes beyond capital cost to include an analysis of operating and maintenance expenses, which is often overlooked in earlier studies. Additionally, the advanced features of ESDs are evaluated. The review's contributions are as follows: It identifies the specific requirements of various ESDs for different application types, with a particular focus on the integration of RESs into the grid. Graphical analyses are conducted to determine the most suitable ESDs for specific services. Through precise identification of application needs and comprehensive assessment of the economic, technical, and environmental implications of storage devices, this review proposes that adopting a hybrid approach to ESDs offers a practical solution for integrating RES and addressing stationary application demands.

These contributions collectively aim to enhance the understanding and deployment of energy storage technologies in supporting a resilient and sustainable energy grid.

- Extensive evaluation of energy storage devices (ESDs) considering technical, economic, and environmental metrics.

- Integration of graphical analysis with evaluation criteria including capital and operational costs.

- Advocacy for hybrid energy storage solutions combining multiple technologies for optimal grid resilience.

- Emphasis on emerging technologies like solid-state batteries and hybrid supercapacitors.

- Focus on ESDs' role in integrating renewable energy sources into the grid, addressing intermittency challenges.

2. State of the art in energy storage devices

The authors conducted an extensive investigation into the requirements of diverse storage applications for grid support and identified suitable ESDs for these purposes. Energy storage plays a multifaceted role across varying power scales to bolster grid operations. At the transmission level, centralized storage systems effectively manage supply and demand fluctuations, spanning from seasonal to hourly variations, crucially stabilizing intermittent energy generation from renewable sources. Centralized storage in transmission also aids in regulating voltage and frequency while enhancing overall efficiency through time-shifting strategies. At lower power levels, such as distribution and consumption, ESDs contribute by offering lower energy densities, thereby managing daily and hourly fluctuations, facilitating peak shaving, and improving system efficiency. The study provides a concise overview of diverse ESD

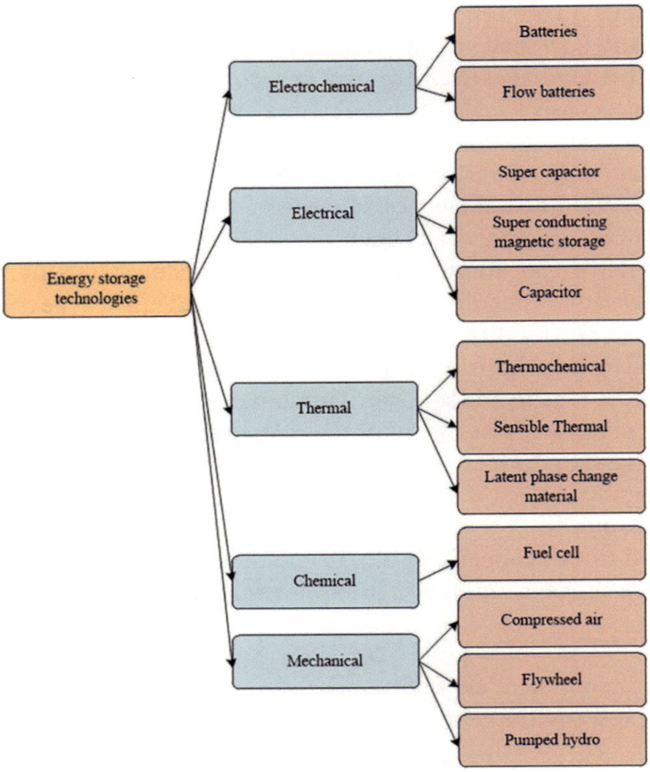

Figure 1.
Classification of various energy storage systems.

types and their respective attributes. A detailed classification of these ESD types is depicted in **Figure 1**.

2.1 Key drivers for energy grid adaptation in the modern energy landscape

The primary drivers necessitating the adjustment of the energy grid in the current energy landscape are multifaceted and interconnected. One of the foremost drivers is the integration of renewable energy sources (RESs) such as wind and solar power. These sources are inherently intermittent and variable, requiring the grid to be more flexible and resilient to manage fluctuations and maintain a reliable power supply. Additionally, the rising energy demand, particularly from electric vehicles (EVs) and other electrified sectors, necessitates a more robust and capable grid to meet the growing consumption needs. The decline in the use of fossil fuels like coal and natural gas, prompted by environmental regulations and economic factors, further underscores the need for the grid to accommodate a larger share of renewable energy sources.

To ensure a continuous and stable energy supply despite the variability of RESs, the grid must incorporate advanced energy storage solutions and other technologies to balance supply and demand effectively. Environmental concerns also play a critical role, as the urgent need to reduce carbon emissions and mitigate climate change impacts drives the shift toward cleaner energy sources and the development of a more sustainable energy grid. Technological advancements in energy storage, such as lithium-ion batteries and thermal energy storage, enable better management

DOI: http://dx.doi.org/10.5772/intechopen.1006499

of energy resources and support the integration of renewable energy into the grid. Economic considerations are equally important, as the cost-effectiveness of integrating advanced energy storage devices and renewable energy technologies into the grid is crucial for long-term sustainability and efficiency. These drivers collectively necessitate significant adjustments to the energy grid to ensure it can support the evolving energy landscape characterized by increased renewable energy integration, rising energy demand, environmental sustainability, and technological advancements.

2.2 Suitability of storage devices for stationary applications

The significant gap in assessing the suitability of different storage devices for various stationary applications can be attributed to several factors. One primary reason is the diverse range of technical, economic, and environmental requirements that different applications entail. Each type of energy storage device, such as lithium-ion batteries, lead-acid batteries, and sodium-sulfur batteries, has unique characteristics that make it suitable for specific contexts but not universally optimal. For instance, while lithium-ion batteries are known for their high energy and power densities, making them suitable for transportation and some stationary uses, their cost and environmental concerns, including the availability of materials like lithium and cobalt, pose significant challenges for broader applications.

Additionally, the evolving energy landscape, driven by the integration of renewable energy sources, necessitates a dynamic adjustment of the energy grid. This complexity is compounded by the intermittent nature of renewable energy, which requires storage solutions that can manage fluctuations in supply and demand across different time scales—from hourly to seasonal variations. Furthermore, the rapid advancement of technology means that new storage solutions are continuously being developed, which adds to the challenge of systematically evaluating their suitability across all potential applications.

Another critical aspect is the varying performance metrics used to evaluate storage devices, which can differ significantly based on the application. Factors such as specific energy, power density, round-trip efficiency, service life, and self-discharge rate are crucial in determining the appropriateness of a storage technology for a particular use case.

2.3 Energy challenges and renewable integration

The projected contributions of wind and solar energy to the global energy demand by 2050 include:

- By 2050, wind and solar energy are expected to fulfill approximately 50% of the world's energy needs.

- This increase is driven by technological advancements, which make these sources more efficient and cost-effective.

- Supportive government policies worldwide are encouraging the adoption of wind and solar power.

- The growth in wind and solar energy will significantly reduce reliance on fossil fuels.

- These renewable energy sources will help decrease greenhouse gas emissions.
- Wind and solar energy will enhance global energy security by diversifying the energy mix.

The reliance on fossil fuels has seen a significant shift since 2018, with a gradual decline as the world transitions toward renewable energy sources. Here are the key points summarizing the changes and projections:

- Decline in fossil fuel usage: Since 2018, there has been a noticeable reduction in the use of fossil fuels such as coal and natural gas. This decline is driven by increasing environmental concerns, regulatory policies, and the economic viability of renewable energy sources.
- Increase in renewable energy: There has been a substantial increase in the adoption of renewable energy sources like wind and solar power. Technological advancements and decreasing costs have made these sources more attractive and feasible.

By 2050, renewable energy sources are projected to dominate the global energy mix, with wind and solar energy alone expected to fulfill approximately 50% of the world's energy needs. This significant shift toward renewables is driven by advancements in technology, decreasing costs, and increasing global efforts to reduce carbon emissions. Meanwhile, the reliance on fossil fuels is anticipated to decline considerably, with forecasts suggesting they will account for about 31% of the total energy mix by 2050. This reduction reflects a growing commitment to transitioning away from high-carbon energy sources. The remaining portion of the energy mix will be composed of contributions from nuclear power, hydro-electricity, and other emerging technologies, each playing a role in supporting a more sustainable and diversified energy portfolio. This diverse energy mix aims to enhance energy security, reduce environmental impacts, and support global economic growth.

2.4 Electrochemical technologies

Li-ion batteries dominate storage installations, comprising more than 85% of new energy storage deployments in 2016. While highly favored, achieving a decarbonized grid necessitates the adoption of diverse energy storage technologies beyond Li-ion batteries alone [5, 8, 10]. Li-ion batteries dominate the field of electrochemical energy storage due to their rapid and extensive adoption across diverse market sectors, ranging from personal electronics to electric vehicles and industrial applications. This widespread usage has led to a significant reduction in the cost of Li-ion battery packs by more than 85% over the past decade [11]. Ongoing research and recent investment trends indicate that new technologies will emerge to better meet daily energy demands with improved reliability and efficiency, potentially driving further cost reductions and increasing deployment in stationary applications. However, challenges persist in integrating batteries into energy storage systems, including the need for longer duration and environmentally friendly solutions. Alternative technologies may offer economically viable and dependable alternatives to address these challenges. Consequently, the suitability of Li-ion storage devices may vary depending on specific requirements for power output and energy density. Evaluation of electric and electrochemical energy storage devices typically considers their performance across specific application contexts [12]. Key technical characteristics of electrochemical energy storage devices are summarized below.

2.4.1 Li-ion batteries

Li-ion batteries are notable for their superior energy and power densities, making them highly suitable for both transportation and stationary uses [13–17]. These attributes are typically quantified using various performance metrics. Key technical specifications of Li-ion batteries include a specific energy range of 75–250 Wh/kg, specific power of 150–315 W/kg, round-trip efficiency between 85 and 95%, a service life spanning 5–15 years, and a self-discharge rate of 0.1–0.3% [18]. Their high specific energy and power densities contribute to their lightweight design, which is advantageous in applications where weight is a critical consideration [17, 18].

They are extensively used in personal electronics, electric vehicles, and industrial applications due to their lightweight design and decreasing costs. However, their limitations in long-duration storage and environmental concerns have led to the exploration of alternative technologies.

Table 2 provides a concise overview of the key technical characteristics and performance metrics of Li-ion batteries.

Characteristic	Description	Performance metrics
Energy density	Gravimetric Energy Density	150–250 Wh/kg
	Volumetric Energy Density	250–700 Wh/L
Efficiency	Coulombic Efficiency	>99%
	Round-Trip Efficiency	85–95%
Cycle life	Number of charge-discharge cycles before capacity significantly degrades	Several hundred to several thousand cycles
Charging and discharging rates	C-Rate: Rate at which a battery is charged or discharged relative to its capacity	0.5–5°C (varies by application)
Temperature range	Effective operating temperature range for maintaining performance and safety	−20 to 60°C
Voltage	Nominal Voltage	3.6–3.7 volts
	Voltage Range	2.5–4.2 volts
Self-discharge rate	The rate at which a battery loses charge when not in use	2–3% per month
Safety features	Mechanisms to prevent overcharging, deep discharging, and overheating	Thermal cut-offs, pressure relief valves, advanced BMS
Environmental impact	Environmental considerations and impact compared to other battery types	Fewer toxic components require careful recycling and disposal handling

Table 2.
An overview of the key technical characteristics and performance metrics of Li-ion batteries.

2.4.2 Lead-acid (Pb-acid) batteries

Lead-acid (Pb-Acid) batteries are recognized for their moderate round-trip efficiency and cost-effectiveness [13, 15, 17]. Typical technical specifications for Pb-Acid batteries include a specific energy range of 30–50 Wh/kg, specific power ranging from 75 to 300W/kg, round-trip efficiency between 70 and 80%, a service life of 5–15 years, and a self-discharge rate of 0.1–0.3% [16–18]. Despite offering comparable service life and self-discharge rates to Li-ion batteries [17, 18], their lower cost makes Pb-Acid batteries particularly suitable for stationary applications [13].

2.4.3 Sodium sulfur (NaS) batteries

Among electrochemical storage devices, Sodium-Sulfur (NaS) batteries are increasingly gaining attention as an emerging technology [12, 17]. NaS batteries are evaluated based on various technical parameters, which include a specific energy range of 150–240 Wh/kg, specific power ranging from 150 to 230 W/kg, round-trip efficiency typically between 80 and 90%, a service life averaging around 15 years, and a negligible self-discharge rate [17, 18]. Notably, NaS batteries are distinguished by their high specific energy, setting them apart from other storage technologies. This characteristic makes them particularly suitable for applications that demand high specific energy [17].

Table 3 outlines the key technical specifications and applications of lead-acid (Pb-Acid) and sodium-sulfur (NaS) batteries, highlighting the differences in energy density, cycle life, efficiency, operating conditions, environmental impact, cost, maintenance requirements, safety, and typical applications.

Specification/Characteristic	Pb-acid batteries	NaS batteries
Energy density	30–50 Wh/kg	150–240 Wh/kg
Cycle life	300–500 cycles	2500–4500 cycles
Efficiency	70–80%	85–90%
Charge/discharge rate	Moderate, typically up to 1°C	High, capable of 2–4°C
Operating temperature range	−20 to 50°C	300–350°C
Self-discharge rate	3–5% per month	<1% per day
Environmental impact	Contains toxic lead and sulfuric acid, and needs careful disposal	High operating temperatures require thermal insulation
Cost	Low initial cost	Higher initial cost
Maintenance	Requires regular maintenance (water refilling)	Low maintenance
Safety	Risk of acid spills and hydrogen gas release	High operating temperature poses safety concerns
Applications	• Automotive (starter batteries) • Uninterruptible power supplies • Emergency lighting • Grid energy storage	• Grid energy storage • Renewable energy integration • Load leveling and peak shaving

Table 3.
Key technical specifications and applications of Pb-Acid and NaS batteries.

2.4.4 Nickel–cadmium (Ni-Cd) batteries

Nickel-cadmium (Ni–Cd) batteries are important in stationary energy storage applications, known for their relatively higher specific power despite a higher self-discharge rate [13, 14]. Key performance parameters for Ni-Cd batteries include a specific energy range of 50–75 Wh/kg, specific power ranging from 150 to 300 W/kg, round-trip efficiency of approximately 70%, a service life spanning 10–20 years, and a self-discharge rate between 0.03 and 0.6% [13–15].

Nickel-cadmium (Ni-Cd) batteries offer several advantages for stationary energy storage applications. One of their primary benefits is their long cycle life, often exceeding 1500–2000 cycles, making them highly durable and suitable for applications requiring frequent charging and discharging. Additionally, Ni-Cd batteries can deliver high discharge currents, which is beneficial for systems needing a rapid release of energy, such as backup power systems and peak shaving applications. Their ability to perform well across a wide operating temperature range, from −20 to 50°C, adds to their versatility, allowing them to be used in various environmental conditions. Moreover, these batteries are robust and resilient, able to withstand electrical abuse, including overcharging and deep discharging, as well as mechanical stress. This robustness enhances their reliability in harsh conditions. Ni-Cd batteries also require relatively low maintenance compared to lead-acid batteries, with no need for regular electrolyte refilling, adding to their convenience.

However, Ni-Cd batteries come with significant disadvantages. They contain cadmium, a toxic heavy metal that poses substantial environmental and health risks. Disposal and recycling of these batteries require careful handling to prevent contamination, making their environmental impact a major concern. Ni-Cd batteries are also susceptible to the memory effect, where the battery gradually loses its maximum energy capacity if not fully discharged and recharged regularly. This can impact their performance over time. Additionally, Ni-Cd batteries have a higher self-discharge rate compared to other battery types, losing approximately 10% of their charge per month when not in use, which affects their efficiency. The initial cost of Ni-Cd batteries is higher than that of lead-acid batteries, which can be a disadvantage for cost-sensitive applications. Furthermore, due to their cadmium content, Ni-Cd batteries face regulatory restrictions in many regions, limiting their use and availability.

The review highlights several primary electrochemical storage options and their suitable applications, which include:

1. Lithium-ion (Li-ion) batteries:

   - Electric vehicles: Li-ion batteries power a wide range of electric vehicles, from cars to busses, providing the necessary energy density for long-range travel.

   - Grid storage: They are increasingly used for grid energy storage solutions, particularly for balancing supply and demand and integrating renewable energy sources like solar and wind power.

   - Industrial applications: Used in various industrial equipment and machinery requiring reliable and efficient energy storage.

2. Lead-acid (Pb-Acid) batteries:

  - Uninterruptible power supplies: Lead-acid batteries are commonly used in UPS systems to provide backup power during outages.

  - Renewable energy storage: Suitable for storing energy generated from renewable sources, such as solar panels, especially in off-grid systems.

  - Automotive: Traditional starter batteries in internal combustion engine vehicles.

3. Sodium-sulfur (NaS) batteries:

  - Grid-scale energy storage: NaS batteries are ideal for large-scale energy storage solutions, providing high specific energy and long service life, suitable for stabilizing the grid and supporting renewable energy integration.

  - Industrial power backup: Used in industrial settings requiring reliable backup power.

4. Nickel-cadmium (Ni-Cd) batteries:

  - Emergency lighting: Ni-Cd batteries are used in emergency lighting systems due to their reliability and durability.

  - Aviation: Employed in aircraft for emergency power and starting the engines.

  - Railway systems: Used for backup power in railway signaling and control systems.

  - Renewable energy storage: Suitable for small-scale renewable energy applications where robustness and durability are essential.

2.5 Thermal energy storage (TES) technologies

Thermal Energy Storage (TES) systems, designed to store heat energy within insulated enclosures, are currently in the initial phases of commercial adoption. TES technology spans a wide temperature spectrum, from −40 to 400°C, accommodating various methods categorized into low-temperature and high-temperature TES. Low-temperature systems encompass options like auriferous and cryogenic energy storage, while high-temperature systems include sensible, latent, and concrete thermal storage techniques [18]. Cryogenic storage stands out due to its high-power capability and extended discharge durations. In sensible heat storage, the storage medium's specific heat capacity plays a crucial role, directly influencing the storage capacity of the TES system [19]. Water offers limited storage capacity, whereas higher capacities can be achieved with latent-phase change materials (PCMs), which leverage the latent heat of the PCMs [20, 21]. Thermochemical storage technology stores and releases heat or cold as required through diverse chemical reactions. Sensible heat storage is currently accessible on the market, whereas PCMs and thermochemical technologies are primarily in developmental stages. Based on various studies, the performance characteristics of TES technologies encompass power capacities ranging from a few MW to 300 MW and energy capacities between 20 and 140 MWh. These systems offer discharge times ranging from hours to several days, an unlimited cycle life, rapid response times in seconds, efficiencies of 30–60%, and energy densities of 80–250 Wh/kg. Additionally, they provide specific energies of 80–250 Wh/kg, specific powers of 10–30 W/kg, service lives spanning 10–30 years, and self-discharge rates of 0.05–1% [5, 18, 22–24].

One of the leading Thermal Energy Storage (TES) technologies utilizes molten salt for energy storage by harnessing and concentrating solar energy. Molten Salt Energy Storage (MSES) systems, which can serve as both storage media and heat transfer agents, operate at high temperatures (up to 570°C) and employ smaller storage tanks [5]. MSES is notable for its efficient heat transfer capabilities and is a commercially viable technology, unlike many other TES technologies still in early development. Despite its advantages, MSES must handle corrosive molten salts that require specific temperature maintenance to prevent freezing, and it is commonly integrated with concentrating solar power plants. The primary advantage of TES is its low self-discharge rate, making it a cost-effective system, albeit with somewhat reduced cycle efficiency [18, 23]. TES serves diverse applications such as electricity generation, heat engine cycle load shifting, and meeting peak energy demands. Moreover, TES plays a role in environmental conservation by reducing reliance on fossil fuels for heat and cold production [24]. At present, TES is predominantly deployed worldwide to facilitate energy time-shifting and enhance the reliability of renewable energy capacities, highlighting its essential function in integrating sustainable energy sources [25].

Thermal energy storage (TES) technologies are crucial for storing thermal energy for later use, helping balance supply and demand in energy systems. There are several types of TES technologies, each with specific temperature ranges and applications.

Sensible heat storage involves storing thermal energy by raising the temperature of a solid or liquid medium. The amount of energy stored depends on the specific heat capacity of the material and the temperature difference. Common media include water, sand, rocks, concrete, and molten salts. The temperature range for these media varies: water (0–100°C), sand and rocks (up to 300°C), and molten salts (150–600°C). Sensible heat storage is straightforward and cost-effective, making it widely used in applications such as domestic hot water systems, building heating and cooling, and solar thermal power plants. Water is commonly used due to its high specific heat capacity and availability, while molten salts are utilized in concentrating solar power (CSP) plants to store solar energy for electricity generation.

Latent heat storage utilizes phase change materials (PCMs) that absorb or release thermal energy during a phase change, such as from solid to liquid or liquid to gas. Common PCMs include paraffin wax, salt hydrates, fatty acids, and eutectic salts, with temperature ranges as follows: paraffin wax (0–100°C), salt hydrates (30–150°C), and eutectic salts (up to 1000°C). Latent heat storage is highly efficient as it stores large amounts of energy at a constant temperature. This technology is used in building temperature regulation, industrial waste heat recovery, and refrigerated transport. The choice of PCM depends on the application temperature range, with paraffin wax and salt hydrates being popular for moderate-temperature applications, while eutectic salts are used for high-temperature industrial processes.

Thermochemical storage involves reversible chemical reactions to store and release thermal energy, storing energy in the form of chemical bonds. Common media include metal hydrides, ammonia-based systems, silica gel, and zeolites. These media operate within different temperature ranges: metal hydrides (300–500°C), ammonia-based systems (−60 to 100°C), and silica gel/zeolites (100–300°C). Thermochemical storage has a high energy density and can store energy for long periods with minimal losses, making it suitable for seasonal storage and long-duration energy applications. For instance, metal hydrides can store hydrogen for high-temperature applications, ammonia-based systems are used for cooling and refrigeration, and silica gel and zeolites are employed in low to moderate-temperature ranges for dehumidification and industrial drying.

Cryogenic energy storage involves cooling air or other gases to cryogenic temperatures (−150 to −273°C) to store energy in the form of liquid gases, such as liquid air and liquid nitrogen. Cryogenic storage is used for large-scale energy storage solutions, particularly for balancing intermittent renewable energy sources like wind and solar. When energy is needed, the liquid gas is vaporized to drive turbines and generate electricity. This technology is advantageous for its high energy density and ability to provide long-term storage.

2.6 Electrical storage

An alternative approach to energy storage includes utilizing electrical storage technologies like Superconducting Magnetic Energy Storage (SMES), supercapacitors, capacitors, and hybrid supercapacitors.

2.6.1 Superconducting magnetic energy storage (SMES)

Superconducting Magnetic Energy Storage (SMES) systems function by converting electrical energy into a magnetic area stored within a superconducting coil during charging. This involves a cooling process to maintain the coil at low temperatures and reduce energy losses. When discharging, SMES releases the stored energy through a power converter configuration [26]. As a well-developed technology, SMES provides a robust solution characterized by high-power density and is particularly suited for short-term energy storage. It holds promise for enhancing the utility of variable renewable energy sources. SMES systems are noted for their high efficiency, exceptional power density, and low degradation rates. However, they also present several drawbacks, including high costs [27], significant self-discharge rates, environmental concerns related to magnetic effects, and sensitivity to temperature fluctuations. SMES systems typically have power capacities ranging from 0.1 to 10 kW and can store up to 100 MWh of energy. They also feature high-power densities, reaching up to 4000 W/L, specific power ratings ranging from 500 to 2000 W/kg, and service lives lasting more than 20 years [18].

2.6.2 Capacitors and supercapacitors

Capacitors: Capacitors are constructed with two or more metal foils separated by a thin insulator, commonly plastic, ceramic, or glass. In the charging process, energy is stored within the dielectric material using an electrostatic field. Capacitors are primarily used for storing small amounts of energy and are well-suited for applications needing high specific power, such as voltage correction and smoothing, owing to their quick charging ability compared to electrochemical cells. Capacitors come in a variety of capacitances and nominal voltages, and recent advancements have focused on enhancing heat dissipation and increasing power levels. Capacitors are widely used in frequency converters, traction systems, and drives. The advantages of capacitors include their quick charging time and high-power output. However, they are limited by their self-discharge losses, low energy provision, limited capacity, and significant energy dissipation, which are considered drawbacks. Capacitors generally offer power capacities ranging from 200 kW to several MW and can store energy between 0.007 kWh to a few kWh. They discharge rapidly within seconds and have a long lifespan of up to 40 years, with efficiency typically ranging from 60–70%. Capacitors exhibit an energy density of 0.07 Wh/kg, specific energy varying between 0.05 and 5 Wh/kg, and specific power ranging from 3000 to 107 W/kg.

Supercapacitors: Supercapacitors, also referred to as electrical double-layer capacitors (EDLCs) or ultracapacitors, are energy storage devices that consist of two carbon electrodes, a porous membrane separator, and an electrolyte. These capacitors can be housed in cylindrical or prismatic shapes, each affecting their energy capacity, power output, volume, and weight characteristics respectively. They can employ either aqueous or nonaqueous electrolytes and can be arranged in different configurations to meet specific application needs. Owing to their design, supercapacitors exhibit properties of both electrochemical cells and traditional capacitors. They typically offer a power capacity extending to several megawatts, energy storage of a few kilowatt-hours, a discharge duration of several minutes, a cycle life of up to 10,610^6106 cycles, and a lifespan of around 10 years at room temperature. Supercapacitors demonstrate efficiencies ranging from 95–98%, energy densities between 4 and 7 Wh/kg, specific energies from 2.5 to 15 Wh/kg, specific power levels from 500 to \ (10^4\) W/kg, and self-discharge rates between 20 and 40%.

The working principle of an SMES system involves several components: a superconducting coil, a refrigeration system, a power conditioning system, and a cryostat to maintain the low temperatures necessary for superconductivity. When energy is supplied to the system, DC flows through the superconducting coil, generating a magnetic field. The energy is stored in this magnetic field as long as the current continues to flow through the superconducting coil. When energy is needed, the system converts the magnetic field energy back into electrical energy by allowing the current to flow out of the coil through the power conditioning system, which manages the conversion and delivery of power to the grid or load.

One of the key benefits of SMES systems is their high efficiency. Since superconductors have no electrical resistance, there are minimal energy losses during the storage and retrieval processes. This efficiency, often exceeding 95%, makes SMES an attractive option for applications requiring rapid response and high-power output, such as grid stabilization, frequency regulation, and uninterruptible power supply (UPS) systems. SMES systems also have extremely fast response times, often in the milliseconds range, allowing them to quickly absorb and release energy as needed, which is crucial for maintaining grid stability and managing power quality.

However, SMES systems also have significant drawbacks. One of the primary challenges is the need for cryogenic cooling to maintain the superconducting state of the coil. This requires sophisticated and expensive refrigeration systems to keep the superconductors at temperatures close to absolute zero (below −196°C for most superconducting materials). The cost and complexity of these refrigeration systems contribute to the high initial investment and operational costs of SMES technology. Additionally, the materials used for superconducting coils, such as niobium-titanium (NbTi) or high-temperature superconductors (HTS), are expensive and require careful handling and maintenance.

Another drawback is the relatively low energy density of SMES systems compared to other energy storage technologies like batteries or pumped hydro storage. While

SMES systems excel in power density and rapid response, they are less suited for applications requiring long-term energy storage or large-scale energy capacity due to their limited storage duration and high costs. Furthermore, the need for continuous cryogenic cooling means that any interruption in the cooling process can lead to a loss of superconductivity, resulting in significant operational challenges.

2.7 Advanced energy storage technologies currently being developed

Several advanced energy storage technologies are currently being developed to enhance the efficiency, capacity, and integration of renewable energy sources into the grid. These technologies include:

1. Gravity-based energy storage: This technology involves utilizing the potential energy of heavy masses, such as large blocks or weights, that are lifted to a certain height. When energy is required, these masses are released to fall, and the potential energy is converted into electricity using generators. This system is particularly useful for grid energy storage, peak shaving, and integrating renewable energy sources, as it can store excess energy generated during low-demand periods and release it when demand is high.

2. Compressed air energy storage (CAES): CAES systems store energy by compressing air and storing it in underground caverns or large, reinforced containers. When electricity is needed, the compressed air is released, heated, and expanded to drive turbines, generating electricity. This method is effective for large-scale energy storage, ensuring grid stability, and supporting the integration of intermittent renewable energy sources like wind and solar.

3. Liquid air energy storage (LAES): LAES technology involves cooling air to a liquid state and storing it in insulated tanks. When energy demand rises, the liquid air is reheated, causing it to expand and turn into a gas that can drive turbines and generate electricity. This system is ideal for large-scale energy storage, providing backup power, and delivering grid services by balancing supply and demand.

4. Hybrid supercapacitors: Hybrid supercapacitors combine the features of supercapacitors and batteries, providing both high-power density and moderate energy density. This combination makes them suitable for applications that require quick bursts of energy as well as steady power output. They are used in grid stability, electric vehicles, and consumer electronics, benefiting from rapid charge and discharge capabilities.

5. High-temperature thermal energy storage: This technology involves storing energy in high-temperature ceramic-based systems or using phase change materials (PCMs) with high melting points. These systems can efficiently store thermal energy for industrial heat processes, concentrated solar power (CSP) plants, and district heating applications, allowing for the utilization of stored heat energy when needed.

6. Pumped thermal energy storage (PTES): PTES systems store energy by transferring heat between two thermal reservoirs, one hot and one cold. The stored

thermal energy is then converted back into electricity using heat engines. This technology is well-suited for grid-scale energy storage, helping balance the supply and demand of renewable energy sources, and providing flexibility in energy management.

7. Iron-air batteries: Iron-air batteries use iron and air as the primary materials, making them a low-cost and environmentally friendly option for large-scale energy storage. These batteries are particularly suitable for grid energy storage, renewable energy integration, and providing backup power, offering a sustainable and cost-effective alternative to traditional batteries.

8. Phase change material (PCM) Batteries: PCM batteries use materials that change phases, such as from solid to liquid, to store and release energy efficiently. These batteries are used for thermal management in buildings, industrial heat storage, and grid energy storage, providing a reliable way to manage temperature fluctuations and store energy in various applications.

These advanced energy storage technologies are being developed to address the limitations of current systems, such as energy density, cost, efficiency, and environmental impact, thereby supporting the transition to a more sustainable and reliable energy infrastructure.

2.8 Optimal energy storage solutions for large-scale power management: key devices and applications

Table 4 highlights the key energy storage devices suitable for large-scale power management, along with their reasons for suitability and primary applications.

Energy storage device	Why suitable for large-scale power management	Applications
Lithium-ion (Li-ion) batteries	High energy density, efficiency, and scalability. Fast response times.	Grid storage, balancing supply and demand, peak shaving, renewable energy integration.
Sodium-sulfur (NaS) batteries	High energy density and long service life. Scalable for large-scale applications.	Grid stabilization, large-scale renewable energy integration.
Flow batteries	Decoupled energy storage capacity and power output. Long cycle life.	Renewable energy storage, grid applications.
Pumped hydro storage (PHS)	High capacity and efficiency for long-duration storage. Established and reliable.	Grid stability, peak load management, storing excess renewable energy.
Compressed air energy storage (CAES)	Large energy storage for extended periods. Effective for grid stability.	Grid stability, integrating intermittent renewable energy sources.
Thermal energy storage (TES)	Effective for storing large amounts of energy in the form of heat. Cost-effective.	Concentrating solar power (CSP) plants, industrial applications, and large-scale energy storage.

Table 4.
Key energy storage devices suitable for large-scale power management, along with their reasons for suitability and primary applications.

2.9 Function of energy storage in the energy supply chain

2.9.1 Crucial role of energy storage in balancing supply and demand across the energy network

Energy storage plays a crucial role in balancing supply and demand at various levels of the energy network by providing flexibility, reliability, and stability. Firstly, one of the primary functions of energy storage systems (ESS) is to enhance grid stability and reliability. ESS can respond quickly to short-term fluctuations in energy supply and demand, helping to stabilize the grid and prevent blackouts. Additionally, they can provide frequency regulation by absorbing excess energy when supply exceeds demand and releasing energy when demand exceeds supply, thus maintaining the grid's frequency within safe limits.

Renewable energy integration is another critical area where energy storage is indispensable. Renewable energy sources such as solar and wind are intermittent by nature. Energy storage systems can store excess energy generated during periods of high production and release it during periods of low production, ensuring a continuous and reliable energy supply. This capability is essential for peak shaving and load leveling, where ESS can reduce the strain on the grid during peak demand periods by discharging stored energy and storing energy during off-peak periods, leveling the load, and optimizing grid performance.

Also, energy storage can help defer infrastructure upgrades. By strategically placing ESS at congested points in the grid, utilities can defer costly infrastructure upgrades. Energy storage systems can manage local supply and demand, reducing the need for new transmission and distribution lines. Additionally, ESS can provide voltage support by absorbing or injecting reactive power, improving power quality, and reducing losses in the transmission and distribution networks.

In microgrids, energy storage plays a vital role. ESS enables island mode operation in microgrids, allowing them to operate independently of the main grid during outages and disturbances, thereby enhancing energy security and resilience. Moreover, ESS can optimize the use of distributed energy resources, such as rooftop solar panels and small wind turbines, by storing excess generation and ensuring that energy is available when needed.

Energy storage systems also facilitate energy arbitrage, which can lead to significant cost savings. ESS can be used for energy arbitrage by buying energy when prices are low (usually during off-peak hours) and selling it or using it when prices are high (during peak hours), leading to cost savings and improved economic efficiency. This practice not only benefits the energy providers but also the consumers by potentially lowering energy costs.

Lastly, ESS provides essential emergency backup power. In the event of emergencies or grid outages, energy storage systems can ensure that critical infrastructure like hospitals, data centers, and communication networks remain operational. This capability is vital for maintaining the functionality of essential services and ensuring public safety during power disruptions.

2.9.2 Role of storage systems in voltage and frequency regulation

For voltage regulation, ESS provides reactive power support, which is essential for stabilizing voltage levels across the grid. They help mitigate voltage drops and sags by supplying or absorbing reactive power, ensuring consistent voltage levels.

Additionally, ESS can manage localized voltage control, particularly in areas with significant fluctuations due to varying demand or renewable energy generation. They also address voltage sags by quickly injecting power during short-term drops and manage voltage swells by absorbing excess power during periods of high generation, preventing harmful voltage rises.

In terms of frequency regulation, ESS offers both immediate and sustained support. For primary frequency response, they can rapidly inject or absorb power to counteract deviations in grid frequency, thus balancing supply and demand in real time. For secondary frequency control, ESS provides longer-term support by continuously adjusting its output to maintain frequency stability until slower-responding resources can take over. Moreover, ESS can participate in frequency regulation markets, offering ancillary services that help grid operators maintain precise frequency control.

Modern ESS are equipped with advanced control systems that allow for real-time adjustments, enabling them to simultaneously manage both voltage and frequency. This dynamic response is particularly valuable for integrating renewable energy sources, which are inherently variable. By smoothing out the intermittency of renewables, ESS ensures that both voltage and frequency remain stable, supporting the increasing share of renewable energy in the grid.

3. Comparison results of the energy storage devices

Section 3 outlines the key performance characteristics of electrochemical, electrical, thermal, and, to some extent, mechanical energy storage technologies. This section, through figures and matrices, underscores that no single energy storage technology currently meets all the requirements for power system applications. Typically, these requirements determine the comparative criteria used for various cases discussed in the preceding sections. To illustrate, the volumetric energy and power capacity of ESDs are depicted in the graphical comparative analysis.

A comparison of thermal energy storage (TES) solutions and electrochemical storage options in terms of capacity and application suitability is presented in **Table 5**.

Table 6 provides a comprehensive overview of the criteria used to evaluate the suitability of different energy storage devices for grid applications. Each criterion is described in detail, with various storage technologies that excel in those areas, highlighting the diverse requirements and capabilities necessary to support a reliable and efficient energy grid.

3.1 Graphical result analysis and selection of ESDs

The assessment of ESDs is conducted primarily through technical, economic, and environmental lenses. To effectively assess the suitability of these ESDs for large-scale grid applications, comprehensive criteria integrating technical, economic, and environmental considerations are employed, utilizing up-to-date and credible data sources. This evaluation is supported by graphical analyses and comparisons. The tables and graphs provided herein reflect mean values obtained from the latest peer-reviewed literature, online sources, and datasheets. These comparisons highlight the diverse metrics of different ESDs. Key technical aspects considered for evaluating selected ESDs include specific energy, specific power, round-trip efficiency, discharge duration, lifetime, and daily self-discharge, detailed in **Table 7**. **Figures 2–4** depict

Aspect	Thermal energy storage (TES)	Electrochemical storage options
Storage Capacity	High	Varies by type; generally lower than TES
Energy Density	Lower	Higher (especially for lithium-ion batteries)
Efficiency	Lower round-trip efficiency due to conversion losses	Higher round-trip efficiency
Cost	Cost-effective for large-scale, long-term storage	Generally higher, though decreasing for technologies like Li-ion
Scalability	Highly scalable for large-scale applications	Scalable, but limited by cost and energy density
Response Time	Slower	Rapid response times
Durability and Lifecycle	Long-term with lower operational costs	Varies; some like lead acid has shorter lifecycles
Primary Applications	Grid stabilization, renewable energy integration, industrial thermal management	Electric vehicles, portable electronics, grid services (peak shaving, load leveling, frequency regulation)
Examples of Technologies	Molten salt storage, ice storage, phase change materials	Lithium-ion, lead-acid, sodium-sulfur, nickel-cadmium batteries
Advantages	Suitable for long-duration, bulk energy storage; lower operational costs	High energy density; efficient for quick discharge/recharge applications
Disadvantages	Lower efficiency due to energy conversion losses	Higher cost; lifecycle and disposal concerns

Table 5.
Comparison of TES solutions and electrochemical storage options in terms of capacity and application suitability.

Criteria	Description	Storage technology
Energy density	The amount of energy stored per unit volume or mass. Higher energy density means more energy can be stored in a smaller space.	Li-ion batteries, Flywheels
Power density	The rate at which energy can be delivered or absorbed per unit volume or mass. Important for applications requiring quick response.	Supercapacitors, SMES
Efficiency	The ratio of the energy output to the energy input is typically expressed as a percentage. Higher efficiency means less energy is lost during storage and retrieval.	Li-ion batteries, SMES
Response time	The time it takes for the storage device to respond to a demand for power. Critical for grid stability and frequency regulation.	Supercapacitors, SMES, Flywheels
Cycle life	The number of charge and discharge cycles a storage device can undergo before its capacity significantly degrades. Longer cycle life means longer service life and lower replacement costs.	Li-ion batteries, Flow batteries

Criteria	Description	Storage technology
Scalability	The ability to increase storage capacity and power output to meet larger demands. Important for accommodating growing energy needs.	Pumped Hydro Storage, Compressed Air Energy Storage (CAES)
Cost	The total cost of ownership, including initial capital costs, operational and maintenance costs, and replacement costs. Lower costs improve economic feasibility.	Lead-acid batteries, Pumped Hydro Storage
Environmental impact	The environmental footprint of the storage device, including resource extraction, manufacturing, operation, and disposal. The lower impact is better for sustainability.	Flow batteries, Pumped Hydro Storage
Safety	The inherent safety risks associated with the storage device, include the potential for fires, explosions, and toxic emissions. Safer technologies are preferred for widespread use.	Lead-acid batteries, Flow batteries
Operational flexibility	The ability of the storage device to perform under various conditions and integrate with other energy systems. Flexible systems can support diverse grid applications.	Li-ion batteries, Flow batteries
Durability and reliability	The robustness of the storage device under operational stresses and its ability to perform consistently over time. Reliable systems ensure continuous power availability.	Flywheels, SMES
Discharge duration	The length of time the storage device can deliver power at its rated capacity. Longer durations are needed for applications like load shifting and backup power.	Pumped Hydro Storage, Flow batteries
Geographic constraints	The limitations imposed by the location and physical space required for the storage system. Some technologies require specific geographic features to be feasible.	Pumped Hydro Storage, CAES

Table 6.
Comprehensive overview of the criteria used to evaluate the suitability of different energy storage devices.

the technical attributes of various ESDs, utilizing average parameter values from **Table 2** for analysis.

Figure 2 illustrates that the SCES system offers the highest specific power among the various ESDs evaluated. Conversely, NaS, Li-ion, PCM, and TCS systems exhibit superior specific energy values. In contrast, PSB and other thermal storage technologies display the lowest specific power metrics. Moreover, both SCES and SMES systems demonstrate the lowest specific energy values.

It is observed that SCES, SMES, NaNiCl2, and Li-ion batteries exhibit very high round-trip efficiencies exceeding 85%. In contrast, STES and Ni-MH systems have a lower cycle efficiency. Regarding service life, TCS stands out with the longest average lifespan of approximately 35 years, whereas electrochemical energy storage devices, such as batteries, typically have shorter service lifespans ranging from 7.67 to 14 years.

Figure 4 shows the daily self-discharge characteristics of different ESDs. $NaNiCl_2$, Ni-MH, and SCES display the highest self-discharge rates, whereas PSB, VRFB, and

Category of energy storage devices	Specific energy (Wh/kg)	Specific power (W/kg)	Round trip efficiency (%)	Service life (years)	Daily self-discharge rate (%)
NaS	150–240 [28, 29] 100–175 [30]	150–230 [28, 29]	75–90 [28, 31] 77 [32]	10–15 [28] 10–20 [30]	20 [28] 0.5–20 [30]
$NaNiCl_2$	120 [31] 100–120 [28, 29] 119 [33]	150 [28, 31] 150–200 [28, 29] 169 [33]	85–90 [29] 92.5 [34]	10–14 [28, 29] >8 [35]	15 [28, 29] 11.887 [33]
Pb-Acid	30–50 [31, 36] 25–32 [37, 38] 20–35 [32, 39]	75–300 [28, 29] 180–200 [30] 25 [32] 74–415 [40]	65–80 [31] 70–90 [28] 85 [41] 72–76 [37]	5–15 [28, 30]	0.1–0.3 [28, 31] Low [32] 0–0.6 general battery [42]
Li-ion	200 [31] 56–200 [28, 29] 90–151 [37] 61–200 [30]	135–315 [28, 29] 61–200 [43] 185–351 [30] 300 [44]	95 [31] 61–90 [35] 59–88 [30] 85 [45]	5–15 [28, 29] 14–16 [30] 6–20 [40]	0.1–0.3 [28, 29] Medium [32] 0.17–0.33 [46] 0.036–0.0833 [47]
Ni-Cd	35–56 [28, 29, 31] 30–61 [30] 40–44 [32, 46]	135–300 [28, 29] 100–144 [30] 140–161 [32] 135 [46]	53 [30] 44–57 [37] 44–51 [29] 51–90 [46]	10–20 [28] 13–20 [30] 10 [40] 5–20 [29]	0.2–0.6 [28] 0–0.6 general battery [42] 0.2–0.3 [30] 0.33 [46]
Ni-MH	44–61 [32] 31–51 [40] 30.13 [48] 51 [49] 89 [50]	200 [49] 220 [32] 158 [50] 440 [44]	35–61 [32] 50 [35, 46]	5–15 [40] >3 [50]	0–0.6 general battery [42] 0.83 [50] 0.3 [50]
VRFB	10–20 [35] 20–29 [37] 20 [30] 25 [32]	150 [30] 61–135 [32]	61 [32, 45] 51–85 [46] 44–85 [29]	10–20 [30, 35] 5–15 [29]	Small [28] Very low [30] 0.2 [29]
PSB	20–29 [37] 10–15 [51]	1.31 [52]	53–83 [37] 44–56 [30] 56 [45]	15 [30, 53] 10–15 [29]	Small [28] 0 [29]
Zn Br	20–29 [37] 34.4–39 [35] 34.4 [54]	90–110 [55]	51 [35] 56 [31] 53–83 [37] 49–85 [30]	8–10 [30] 5–10 [29]	Small [28] 0.24 [29]
SCES	2.5–15 [29] 0.5–1.5 [28] 11 [56] 4 [57]	10,000 [31] 350–3500 [28, 29] 2000–3500 [30] 610–2000 [58]	95 [31, 32] 95–99 [37] 49–90 [30]	20 [28, 29] 8–17 [30]	5–40 [31] 20–40 [28, 29]
SMES	10–56 [30] 1–12 [59] 3 [59]	350–2000 [28, 29]	95 [58] 61–95 [30] 90–98 [29]	20 [30] 20–30 [29, 57]	10–15 [28, 29]
STES	61–120 [28] 10–35 [60] 61–235 [24]	10–30 [24]	35–90 [61] 44 [58] 35–90 [60] 30–44 [24]	10–20 [28] 10–30 [24]	0.5 [28] 0.05–1 [24]

Category of energy storage devices	Specific energy (Wh/kg)	Specific power (W/kg)	Round trip efficiency (%)	Service life (years)	Daily self-discharge rate (%)
PCM	135–235 [28] 61–235 [24]	10–30 [28]	56–90 [61] 35–90 [60] 40–90 [41]	20–40 [28] 30 [41]	0.5–1 [28]
TCS	235 [60] 61–235 [24]	10–30 [24]	56–100 [61] 56–100 [60] 40–90 [41] 30–44 [24]	10–30 [24] 30 [41]	0.05–1 [24]

Table 7.
Key technical characteristics of diverse energy storage technologies.

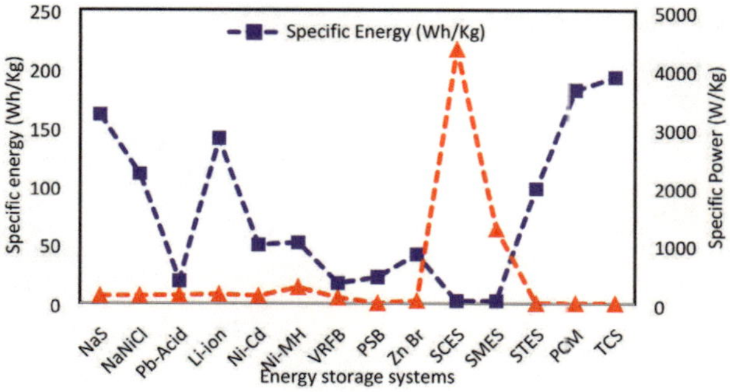

Figure 2.
*Comparison of specific energy and specific power of ESD based on the average values derived from **Table 2**.*

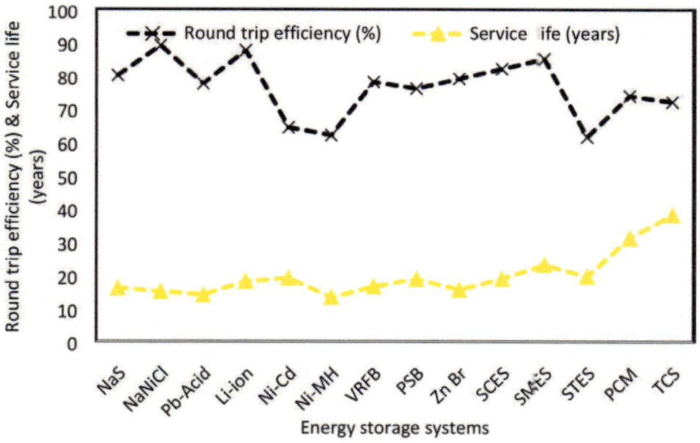

Figure 3.
*Mean cycle efficiency and operational longevity of energy storage systems as detailed in **Table 2**.*

Li-ion technologies exhibit lower self-discharge ratios. Particularly notable among electrochemical ESDs, $NaNiCl_2$ and NaS demonstrate considerable daily self-discharge rates. **Figure 5** demonstrates how the size of energy storage devices (ESDs) correlates with their power and energy densities. Higher energy and power densities result in smaller ESD volumes, shown in the top right corner of the graph. Conversely,

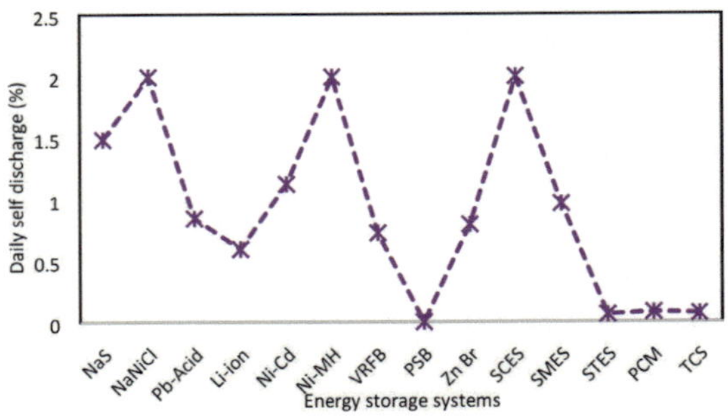

Figure 4.
Typical daily self-discharge rates of energy storage devices from data presented in ***Table*** *2.*

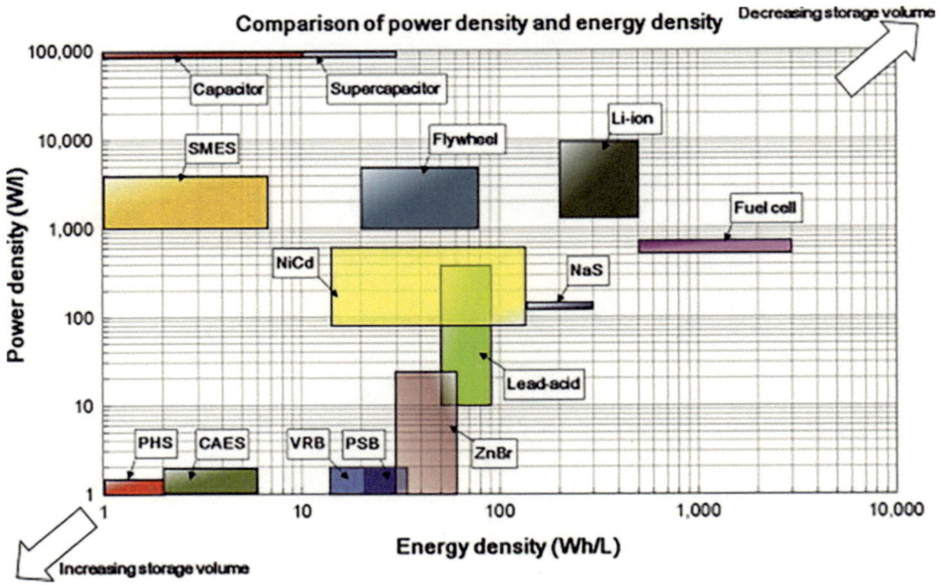

Figure 5.
Comparative analysis of power and energy densities across energy storage devices.

ESDs with lower energy and power densities occupy larger volumes, as indicated in the bottom left corner. Thermal ESDs and electrochemical ESDs like Ni-MH, Na-S, Li-ion, and $NaNiCl_2$ generally exhibit higher energy densities. In contrast, SCES, SMES, and FES show higher power densities. CAES and PHS have the lowest energy densities, while CAES, PHS, VRFB, PSB, and Zn-Br have the lowest power densities. Li-ion batteries are notable for combining high energy and power densities, resulting in compact sizes suitable for diverse applications, including portable devices, transportation, and grid-scale systems.

Figure 6, depicted as a bubble chart, compares discharge times and power ratings among various ESDs, revealing distinctive characteristics. Mechanical energy storage systems such as CAES and PHS demonstrate extended discharge durations and higher power ranges compared to other technologies. In contrast, SCES, FES, and SMES are characterized by short discharge times and lower power ranges.

DOI: http://dx.doi.org/10.5772/intechopen.1006499

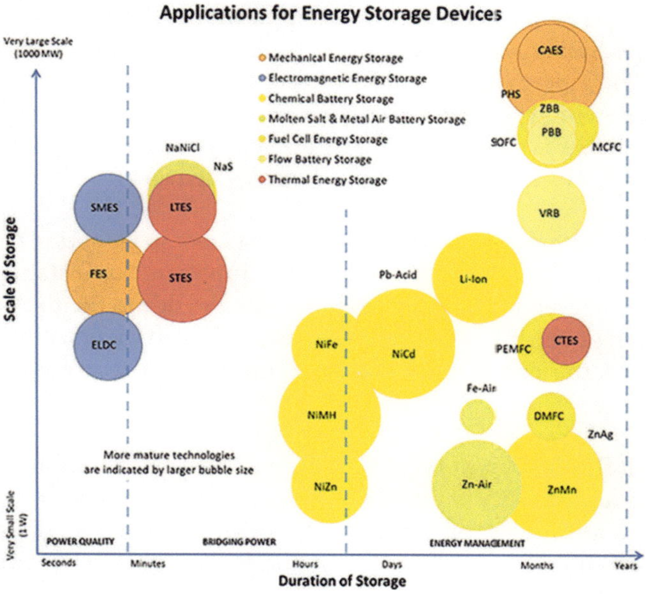

Figure 6.
Comparison of ESDs based on average discharge time and power rating.

3.2 Future outlook and hybrid solutions

The review suggests using a mix of technologies in hybrid solutions for energy storage to leverage the unique strengths of each technology while mitigating their individual limitations. Different energy storage technologies excel in various aspects, such as power density, energy density, response time, cycle life, and cost. For instance, while some technologies like supercapacitors provide high power density and quick response times, they may not offer sufficient energy density for long-duration storage. Conversely, technologies such as lithium-ion batteries or thermochemical storage provide high energy density, suitable for prolonged energy discharge but may lack the rapid discharge capabilities required for certain applications.

By integrating multiple technologies in a hybrid system, it is possible to create a more versatile and efficient energy storage solution that can handle a wide range of applications and operational conditions. For example, a hybrid system combining supercapacitors and lithium-ion batteries can provide both immediate power response and sustained energy supply, thus supporting both grid stability and long-term energy storage needs. This approach allows for optimization of the overall system performance, cost-effectiveness, and operational flexibility, making hybrid energy storage solutions particularly valuable in complex and variable energy markets. Additionally, using a mix of technologies can enhance the reliability and resilience of the energy storage system, ensuring continuous power supply even when one component experiences issues.

Hybrid energy storage solutions are designed to combine multiple energy storage technologies to address the diverse and unique requirements of different energy applications. Each energy storage technology has its own strengths and limitations; by integrating them into a hybrid system, it is possible to harness the advantages of each while compensating for their weaknesses.

1. Power and energy density needs: Some applications require high-power density (fast charge/discharge rates) while others need high energy density (long-duration storage). Supercapacitors or flywheels, which offer high-power density, can be combined with lithium-ion batteries, which provide high energy density, to cater to applications like frequency regulation (which needs quick response) and load leveling (which requires sustained energy delivery).

2. Cost and efficiency optimization: Hybrid systems can be optimized to use the most cost-effective technology for a specific function. For instance, flow batteries might be used for large-scale energy storage due to their relatively lower costs and long cycle life, while more expensive but efficient technologies like lithium-ion batteries are used for applications needing high energy density and fast response.

3. Operational flexibility and reliability: Hybrid systems can improve the overall reliability and flexibility of energy storage. By integrating different technologies, the system can continue to operate effectively even if one component fails or underperforms. This is crucial for maintaining grid stability and ensuring a continuous power supply.

4. Environmental and space considerations: Different technologies have different environmental impacts and space requirements. For instance, pumped hydro storage requires specific geographic features and large areas, while batteries can be installed in more diverse locations. Hybrid systems can optimize space use and minimize environmental impacts.

5. Application-specific requirements: Different applications, such as residential energy storage, industrial use, or renewable energy integration, have varying requirements in terms of scale, efficiency, and performance. Hybrid solutions can be tailored to meet these specific needs more precisely than a single technology might.

4. Conclusion

Energy storage plays a pivotal role in the future electricity grid, particularly in achieving the goal of generating 70% of electricity from RESs. It offers flexibility to balance supply and demand and facilitates the efficient integration of RESs. As a result, there is an anticipated increase in energy storage capacity, driving the development of a diverse array of electric and thermal power technologies tailored to meet operational and economic requirements. This comprehensive review examines the maturity, current status, and future directions of various storage technologies, with a primary focus on electrical, electrochemical, and thermal storage systems. The key findings highlight the advancements and potential of ESDs.

- Considerations of source availability, accessibility, and environmental impact are crucial for the life cycle analysis of electrochemical energy storage systems. For instance, producing 1 kWh of lithium-ion battery requires approximately 400 kWh of energy and emits 75 kg of CO_2, whereas a coal-fired plant emits 1 kg of CO_2 per kWh. This highlights the necessity for lithium-ion systems to undergo at least 400 cycles to offset their energy consumption and achieve a positive environmental impact over their long service life. Among electrochemical ESDs, lithium-ion batteries stand out for their higher power and energy densities, superior round-trip efficiency, minimal environmental footprint, and lightweight characteristics. Consequently, they are regarded as a promising choice for grid-scale stationary applications, particularly in integrating renewable energy sources (RESs) into the grid.

- The availability of lithium and cobalt poses challenges to the current technology of Li-ion cells in terms of cost and scalability. Additionally, electrodes often contain toxic elements such as lead, cadmium, and mercury, which raise significant health and environmental concerns. Addressing these challenges through practices like recycling, second-life utilization, and innovative concepts such as Vehicle-to-Grid (V2G) and Grid-to-Vehicle (G2V) can play a crucial role. These approaches are essential for mitigating environmental impacts and overcoming the limitations associated with these materials.

- Utilizing TES derived from renewable sources offers substantial potential to decrease CO_2 emissions across residential, nonresidential, and industrial sectors by conserving significant amounts of energy. However, TES encounters challenges related to costs and stability, particularly with emerging technologies such as TCS and PCMs. Similar to other energy storage solutions, designing TES systems must be tailored to meet specific application boundaries and requirements.

Author details

Sina Samadi Gharehveran[1]*, Kimia Shirini[1] and Arya Abdolahi[2]

1 Faculty of Electrical and Computer Engineering, University of Tabriz, Tabriz, Iran

2 Azarbaijan Shahid Madani University, Tabriz, Iran

*Address all correspondence to: s.samadi@tabrizu.ac.ir

References

[1] Ali U. BN4.EF. New Energy Outlook. Power Technology. 27 Jun 2019. [Online]. Available from: https://www.power-technology.com/news/bloomberg-new-energy-outlook-2019-2/

[2] BNEF. Electric Vehicle Outlook. 1 Apr 2020. [Online]. Available from: https://about.bnef.com/electric-vehicle-outlook-2020/

[3] BNEF. Energy Storage Investments Boom as Battery Costs Halve in the Next Decade. 2019. Available from: http://seo.org.pl/en/bnef-do-2030-ceny-baterii-spadna-o-polowe/ [Accessed: May 2, 2021]

[4] European Commission, Andrey C (Artelys), Barberi P (Artelys), Lacombe L (Artelys), van Nuffel L (Trinomics), G'erard F (Trinomics), et al. In: (Trinomics), El Idrissi Y (enerdata), Morgan Crenes (enerdata). Study on Energy Storage-Contribution to the Security of the Electricity Supply in Europe; 2020. DOI: 10.2833/077257

[5] Balducci M, Bridges M, Engel P, Lindstrom A, Liu P, Santhiyakumari A, et al. Five Steps to Energy Storage: Deployment at Federal Sites. US: Department of Energy; Aug 2020. [Online]. Available from: https://www.energy.gov

[6] Rastler D. Electricity Energy Storage Technology Options: A white Paper Primer on Applications, Costs, and Benefits. Palo Alto, CA: Electric Power Research Institute (EPRI); 2010

[7] Akhil AA, Huff G, Currier AB, Kaun BC, Rastler DM, Chen SB, et al. DOE/EPRI 2013 Electricity Storage Handbook in Collaboration with NRECA. Albuquerque, NM, USA: Sandia National Laboratories; Jul 2013

[8] Hemmati R, Saboori H. Emergence of hybrid energy storage systems in renewable energy and transport applications—A review. Renewable and Sustainable Energy Reviews. 2016;**65**:11-23

[9] European Commission. The Future Role and Challenges of Energy Storage. DG ENER Working Paper; 2013. pp. 1-36. Available from: https://ec.europa.eu/energy/sites/ener/files/energy_storage.pdf

[10] IEA. Technology mix in storage installations excluding pumped hydro. Paris: IEA; 2011-2016. [Online]. Available: https://www.iea.org/data-and-statistics/charts/technology-mix-in-storage-installations-excluding-pumped-hydro-2011-2016

[11] Thielmann A, Sauer A, Isenmann R, Wietschel M. Technology Roadmap Energy Storage for Electric Mobility 2030. Karlsruhe, Germany: Fraunhofer Institute for Systems and Innovation Research; 2013. p. 32

[12] Maddukuri S, Malka D, Chae MS, Elias Y, Luski S, Aurbach D. On the challenge of large energy storage by electrochemical devices. Electrochimica Acta. 2020;**354**:136771

[13] Cole WJ, Frazier A. Cost projections for utility-scale battery storage. National Renewable Energy Lab.(NREL), Golden, CO (United States); 19 Jun 2019

[14] Goldie-Scot L. A behind the scenes take on lithium-ion battery prices. 2019. [Online]. Available from: https://about.bnef.com/blog/behind-scenes-take-lithium-ion-battery-prices/

[15] Pillot C. The rechargeable battery market and main trends 2018-2030.

In: 36th Annual International Battery Seminar & Exhibit. Avicenna Energy. Mar 2019. p. 35

[16] Cruz MRM, Fitiwi DZ, Santos SF, Catal˜ao JPS. A comprehensive survey of flexibility options for supporting the low-carbon energy future. Renewable and Sustainable Energy Reviews. 2018;**97**:338-353

[17] Salkuti SR. Energy storage technologies for smart grid: A comprehensive review. Majlesi Journal of Electrical Engineering. 2020;**14**:39-48

[18] Luo X, Wang J, Dooner M, Clarke J. Overview of current development in electrical energy storage technologies and the application potential in power system operation. Applied Energy. 2015;**137**:511-536

[19] Koçak B, Fernandez AI, Paksoy H. Review on sensible thermal energy storage for industrial solar applications and sustainability aspects. Solar Energy. 2020;**209**:135-169

[20] Nie B, Palacios A, Zou B, Liu J, Zhang T, Li Y. Review on phase change materials for cold thermal energy storage applications. Renewable and Sustainable Energy Reviews. 2020;**134**:110340

[21] Gholamibozanjani G, Farid M. A comparison between passive and active PCM systems applied to buildings. Renewable Energy. 2020;**162**:112-123. DOI: 10.1016/j.renene.2020.08.007

[22] Gibb D, Seitz A, Johnson M, Romani J, Gasia J, Cabeza LF, Gurtner R. Applications of Thermal Energy Storage in the Energy Transition. 2018. p. 154. Available from: https://www.eces-a30.org/wp-content/uploads/Applications-of-Thermal-Energy-Storag-in-the-Energy-Transition-Annex-30-Report.pdf

[23] Aziz MB, Zain ZM, Baki SR, Muslam MN. Review on performance of thermal energy storage system at S&T complex, UiTM Shah Alam, Selangor. In: 2010 IEEE Control and System Graduate Research Colloquium (ICSGRC 2010). Piscataway, NJ, USA: IEEE; 22 Jun 2010. pp. 49-54

[24] IRENA-ETSAP. Thermal Energy Storage Technology Brief E17. 2013. Available from: https://www.coursehero.com/file/16594128/IRENA-ETSAP-Tech-Brief-E17-Thermal-Energy-Storage/ [Accessed: May 6, 2021]

[25] IRENA. Electricity Storage and Renewables: Costs and Markets to 2030. 2017. Available from: https://www.irena.org/publications/2017/oct/electricity-storage-and-renewables-costs-and-markets [Accessed: May 10, 2021]

[26] Ali MH, Wu B, Dougal RA. An overview of SMES applications in power and energy systems. IEEE Transactions on Sustainable Energy. 2010;**1**:38-47

[27] Smith SC, Sen PK, Kroposki B. Advancement of energy storage devices and applications in electrical power system. In: 2008 IEEE General Meeting Power and Energy Society. 2008. pp. 1-8. DOI: 10.1109/PES.2008.4596436

[28] Chen H, Cong TN, Yang W, Tan C, Li Y, Ding Y. Progress in electrical energy storage system: A critical review. Progress in Natural Science. 2009;**19**:291-312

[29] Bradbury K. Energy Storage Technology Review Kyle Bradbury. 2010. Available from: https://www.kylebradbury.org/docs/papers/Energy-Storage-Technology-Review-Kyle-Bradbury-2010.pdf

[30] Díaz-Gonz'alez F, Sumper A, Gomis-Bellmunt O, Villaf'afila-Robles R. A review of energy storage technologies

for wind power applications. Renewable and Sustainable Energy Reviews. 2012;**16**:2154-2171

[31] Beaudin M, Zareipour H, Schellenberglabe A, Rosehart W. Energy storage for mitigating the variability of renewable electricity sources: An updated review. Energy for Sustainable Development. 2010;**14**:302-314

[32] Vazquez S, Lukic SM, Galvan E, Franquelo LG, Carrasco JM. Energy storage systems for transport and grid applications. IEEE Transactions on Industrial Electronics. 2010;**57**:3881-3895

[33] Dustmann CH. Advances in ZEBRA batteries. Journal of Power Sources. 10 Mar 2004;**127**(1-2):85-92

[34] Matheys J, Timmermans JM, Van Mierlo J, Meyer S, Van den Bossche P. Comparison of the environmental impact of five electric vehicle battery technologies using LCA. International Journal of Sustainable Manufacturing. (Geneva, Switzerland: Inder Science Publishers). 1 Jan 2009;**1**(3):318-329

[35] Wikipedia. Lithium Battery. 2013. Available from: https://en.wikipedia.org/wiki/Lithium_battery [Accessed: March 27, 2021]

[36] Mobbs M. Lead acid vs. Lithium-ion battery comparison—Life cycle and performance. Sustainable Projects. 2015;7:567-580. Available from: www.sustainablehouse.com.au http://www.streetcoolers.com.auwww.ecopops.com.au

[37] Tag Electricity Storage Association. Energy Storage World Forum. 2015. Available from: https://energystorageforum.com/tag/electricity-storage-association [Accessed: March 31, 2021]

[38] Das CK, Bass O, Kothapalli G, Mahmoud TS, Habibi D. Overview of energy storage systems in distribution networks: Placement, sizing, operation, and power quality. Renewable and Sustainable Energy Reviews. 2018;**91**:1205-1230

[39] Battery I, Energy A. The Rechargeable Battery Market and Main Trends 2014-2025. 2015. Available from: http://www.avicenne.com/pdf/Fort_Lauderdale_Tutorial_C_Pillot_March2015.pdf

[40] McCluer S, Christin JF. Comparing data center batteries, flywheels, and ultracapacitors. White Paper. No. 65. France: Schneider Electric, Rueil-Malmaison; 2008. p. 202

[41] Palizban O, Kauhaniemi K. Energy storage systems in modern grids—Matrix of technologies and applications. Journal of Energy Storage. 2016;**6**:248-259

[42] Hedegaard K, Meibom P. Wind power impacts and electricity storage—A time scale perspective. Renewable Energy. 2012;**37**:318-324

[43] Arani AK, Karami H, Gharehpetian GB, Hejazi MS. Review of flywheel energy storage systems structures and applications in power systems and microgrids. Renewable and Sustainable Energy Reviews. (Amsterdam, Netherlands: Elsevier). 1 Mar 2017;**69**:9-18

[44] Theoildrum.com. Energy Storage—Flywheel. 2011. Available from: http://theoildrum.com/node/8428 [Accessed: March 28, 2021]

[45] Schaber C, Mazza P, Hammerschlag R. Utility-Scale Storage of Renewable Energy. 2004. pp. 1040-6190. Available from: https://ideas.repec.org/a/eee/jelect/v17y2004i6p21-29.html

[46] Wagner L. Overview of energy storage methods. Analyst. Royal Society of Chemistry, Cambridge. 2007;**132**(12):1313-1326

[47] Kondoh J, Yamaguchi H, Murata A, Otani K, Sauta K, Higuchi N, et al. Electrical energy storage systems for energy networks. Energy Conversion and Management. 2000;**41**:12

[48] Manual A. Nickel Metal Hydride (NiMH). 2018. pp. 1-14. Available from: https://data.energizer.com/pdfs/nickelmetalhydride_appman.pdf

[49] Kopera JJC. Inside the Nickel Metal Hydride Battery inside the NiMH Battery. 2004. Available from: https://citeseerx.ist.psu.edu/viewdoc/download?doi=10.1.1.729.737&rep=rep1&type=pdf

[50] Tan B, Flood S, Jacinto J, Hom D, SOS Beacon. [Ph.D. dissertation]. New Brunswick, NJ, USA: Rutgers University; 2023

[51] Rydh CJ, Sande A. Energy analysis of batteries in photovoltaic systems. Part I: Perform Energy Require. 2005;**46**:1957-1979

[52] Morrissey P. Regenesys: A New Energy Storage Technology. 2016. p. 750. DOI: 10.1080/01430750.2000.9675376

[53] DeBoer P. Flow Batteries. 2007. pp. 1-9. Available from: http://www.leonardo-energy.org-Flow [Accessed: April 5, 2021]

[54] News ES. Zinc Bromine Flow Batteries for Large Scale Stationary Electricity Storage. 2009. Available from: http://www.energystoragenews.com/Zinc-Bromine-Flow-Batteries-for-Large-Scale-Stationary-Electricity-Storage.html [Accessed: February 25, 2021]

[55] Rajarathnam GP. The Zinc/Bromine Flow Battery. 2016. Available from: https://ses.library.usyd.edu.au/handle/2123/16394

[56] Conte M, Prosini PP, Passerini S. Overview of energy/hydrogen storage: State-of-the-art of the technologies and prospects for nanomaterials. Journal of Power Sources. (Amsterdam, Netherlands: Elsevier). 2004;**108**:2-8

[57] Adetokun BB, Oghorada O, Abubakar SJ. Superconducting magnetic energy storage systems: Prospects and challenges for renewable energy applications. Journal of Energy Storage. (Amsterdam, Netherlands: Elsevier). 25 Nov 2022;**55**:105663

[58] Ibrahim H, Ilinca A, Perron J. Energy storage systems—Characteristics and comparisons. Renewable and Sustainable Energy Reviews. 2008;**12**(5):1221-1250

[59] Nadeem F, Hussain SS, Tiwari PK, Goswami AK, Ustun TS. Comparative review of energy storage systems, their roles, and impacts on future power systems. IEEE Access. 2018;**7**:4555-4585

[60] Renewable I, Agency E. IRENA-IEA-ETSAP Technology Brief 4: Thermal Storage. 2013. Available from: http://www.inship.eu/docs/TES%201%20IRENAETSAP%20Tech%20Brief%20E17%20Thermal%20Energy%20Storage.pdf

[61] Sarbu I, Sebarchievici C. A comprehensive review of thermal energy storage. Sustainability. 2018;**10**(1):191. DOI: 10.3390/su10010191

Section 2

Integration Feasibility of Battery Energy Storage Systems

Chapter 3

Techno-Economic Assessment of a Grid-Connected Residential Rooftop Photovoltaic System with Battery Energy Storage System

Ciprian Cristea, Maria Cristea, Radu-Adrian Tîrnovan and Florica Şerban

Abstract

Grid-connected residential rooftop photovoltaic systems with battery energy storage systems are being progressively utilized across the globe to enhance grid stability and provide sustainable electricity supplies. Battery energy storage systems are regarded as a promising solution for overcoming solar energy intermittency and, simultaneously, may reduce energy expenditure by minimizing grid exports or maximizing solar electricity self-consumption by households. This chapter aims to assess the feasibility of six lithium-ion and lead-acid batteries with different capacities connected to a grid-connected rooftop solar photovoltaic system for a dwelling situated in the north-western part of Romania. The results pointed out that the most viable option is the photovoltaic system combined with the 16.8 kWh lead-acid battery, generating an additional value of almost $18,000 USD over the investment's lifetime. The results provide important decision-making information as regards the viability of grid-connected photovoltaic systems combined with different battery energy storage system technologies.

Keywords: battery energy storage system, grid-connected photovoltaic system, renewable energy, solar energy, feasibility analysis, prosumer

1. Introduction

Storage systems represent the key solution to facilitate the integration of renewable energy sources (RES) without causing massive grid disturbances, such as voltage variations and power fluctuations that can affect energy efficiency and increase overcurrent [1]. According to references [2, 3], electrical energy storage systems (EESS) can be grouped into four categories, depending on the form of electrical energy stored: mechanical storage, chemical storage, electromagnetic storage, and thermal storage. The storage technologies included in each category are presented in **Figure 1**.

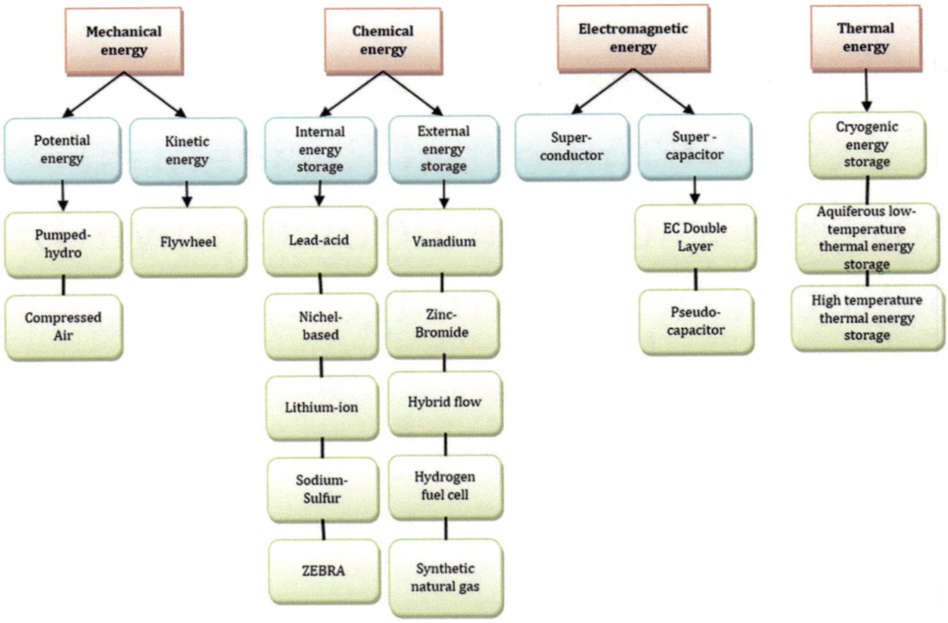

Figure 1.
Technologies used in electrical energy storage [2, 3].

The first category, mechanical energy storage, included the systems based on potential energy, such as pumped hydropower and compressed-air technologies and kinetic energy, used in flywheel storage. It has been discovered that the needed long-duration energy storage capacity is influenced by the renewable energy mix [4]. Also, the long-duration storage technologies may provide significant cost savings in electricity systems that rely substantially on solar and wind power generation [5].

Pumped storage hydropower is the most widely used large-scale technology for high power applications (>100 MW). This type of system is based on pumping water from a lower reservoir to an upper one by using energy during periods with low demand and cost (night, weekends, holidays) and producing electricity during peak periods when the price of electricity is high. They are frequently located near wind and large solar farms [2].

Compressed-air energy storage is considered the most economically efficient and can contribute to the development of sustainable systems used to store energy produced by renewable energy systems. Although they provide large capacities (>100 MW), they require special facilities and large underground storage spaces. Conventional compressed air energy storage systems are associated with greenhouse gas emissions. They show low efficiency (42–54%) due to high heat losses to the atmosphere during compression but also due to complex thermal energy requirements when the air expands in the turbine, causing it to cool [2].

The flywheel inertial system stores kinetic energy in a disk that rotates at a very high speed and is connected to the shaft of an electric machine, injecting the stored energy into the grid [6]. Its advantages are high efficiency 90–95%, extended duty cycle life, lack of sensitivity to ambient conditions, and no releasing of hazardous chemicals. However, they fail to compensate for the main drawback—the self-discharge performance is almost 100% daily, which is why they are not used for long-term storage [2]. The applications in which this storage technology is used are [6]—hybrid vehicles, railway applications, wind systems, marine, and space applications.

Chemical energy can be obtained in internal and external energy storage. Internal storage batteries consist of cells containing a pair of electrodes of opposite polarity and immersed in an electrolyte. It uses electrochemical reactions to store and supply electricity at desired times. Depending on the materials used for electrodes and electrolytes, batteries are divided into lead-acid, nickel-cadmium and nickel-metal-hydride, metal-air, lithium-ion, and sodium-sulfur battery, respectively, sodium-metal-chloride (ZEBRA) [2].

External storage batteries are redox flow batteries and fuel cells. Redox flow batteries are considered more advantageous than conventional ones due to their unlimited storage capacity. The chemical potential obtained from the electrical energy is stored in external reservoirs of different sizes. Although they require additional equipment that generates high costs, redox flow batteries have several advantages: low self-discharge rate, long life, and short response time. Despite their high cost, fuel cells offer the highest specific energy and long service life, which recommends them as a long-term storage system [2].

Chemical energy storage systems are often used in stationary applications, including as storage systems integrated in residential photovoltaic systems. The main batteries are presented in detail in the next section, revealing their technical performances and the advantages and disadvantages of each analyzed technology.

The electromagnetic energy storage systems category consists of superconductors, which involve storing electrical energy in the magnetic field created by current flow in a superconducting coil with low resistance at low temperatures, and supercapacitors, which store electricity in the electrostatic field of the electrochemical double layer, determining numerous charge–discharge cycles and a long lifetime. The applications where they are used include the portable electronics industry, electric or hybrid vehicles to provide power during acceleration, and the space industry to operate satellite radar systems [2].

The last category, thermal energy storage, is represented by cryogenic energy storage, aquifer thermal energy storage, and very high temperature thermal energy storage. These technologies are still in the development phase, their main advantage being the minor impact on the environment [7].

Of all energy storage systems presented, several chemical energy storage systems are often integrated in residential roof-top photovoltaic systems. Thus, these technologies are further analyzed to identify the most viable solution from a technical and economical point of view.

This chapter is structured as follows: in the second section, the battery energy storage systems (BESS) are described in detail, outlining the characteristics and performance of each technology. In the third section, a technical comparison between the analyzed BESS is made to highlight the advantages and disadvantages of each battery. The fourth section evaluates the feasibility of six lithium-ion (Li-ion) and lead-acid (LAB) batteries with various capacities connected to a photovoltaic (PV) system for a dwelling in Romania. In the last section, the conclusions of this chapter are drawn.

2. Battery energy storage system technologies

In recent years, solar PV projects have developed greatly [8]. The installed power in photovoltaic installations has grown in both solar plants and residential PV systems. Thus, the integration of BESS is crucial to ensuring grid stability.

There are several BESS technologies that can be integrated in residential rooftop PV systems. The most widely used are Li-ion, LAB, and lead-carbon (LCB) batteries.

Modern BESS technologies are in the research-development stage, and they are commercialized in limited numbers: vanadium redox flow battery (VRFB), aqueous ion hybrid battery (AIHB), sodium-sulfur battery (NaS), and proton-exchange membrane fuel cells (PEMFC).

2.1 Li-ion batteries

Li-ion batteries are considered the preferred BESS for residential applications due to their technical performance. The main advantages are high specific power (250–340W/kg) and energy density [9], determining a high operating voltage (around 4 V) due to non-aqueous electrolytes [10].

Compared with LAB, Li-ion batteries have higher efficiency (90–99%), longer lifetime over 6000 life cycles depending on operating conditions, lower self-discharge rate (8–31%), and lower environmental impact. Other specific characteristics are low maintenance, long service life, high cell voltage, and lightweight [11].

The disadvantages of Li-ion batteries are related to their operation conditions. When the cells are connected in series or parallel and the voltage range limit is exceeded, exothermal reactions occur, and the battery may explode or burn. Also, the deep discharge process and operating at high temperatures shorten the lifespan of these batteries [12].

Li-ion batteries are also used in other applications, such as laptops, smartphones, machines tools, and electric vehicles. There are various types of lithium-based batteries, depending on the materials used in their composition [13]. A typical Li-ion battery has four components: an anode and a cathode kept apart by a porous separator and by the electrolyte. The anode represents the negative terminal, while the cathode is the positive one and the source of lithium ions, which move through the electrolyte in the anode, where the current is allowed to pass in an external circuit [12]. Usually, graphite is used as an anode, and variants of lithium metal amalgams as cathodes. The main technologies are lithium iron phosphate, lithium-titanate, lithium-sulfur, and lithium-air batteries [13].

2.2 Lead-acid batteries

Lead-acid batteries represent the mature storage technology and the most economically efficient alternative to Li-ion batteries. Their structure is relatively simple and similar to Li-ion batteries—both negative and positive electrodes are immersed in an electrolytic sulfuric acid and water solution, which permits the passage of ions, and are kept apart by a separator. Valve-regulated batteries are sealed, but they have a valve to release the gases when the internal pressure is too high. The design involves a minimal amount of immobilized electrolyte, which allows the reuse of oxygen in the battery, implying less or no water loss [14]. Based on the electrolyte, there are two technologies intensively commercialized: absorptive glass mat valve-regulated lead acid (AGM-VRLA) and gel valve-regulated lead acid (Gel-VRLA), each one with its own structure and characteristics [11].

AGM-VRLA technology proprieties depend on material porosity and uniformity. The electrodes are separated by an absorptive glass mat, which also retains through wetting and wicking the sulfuric acid, making it safe to use. The electrolyte distribution depends on porous structure and the saturation level can affect recharge and discharge performances [15]. The main characteristics are high power, fast charging capabilities, and low maintenance requirements. Despite their low energy density, low efficiency between 80% and 90%, and few life cycles, 500–2500 at 30–50% depth of discharge (DoD) [11], AGM-VRLA batteries are frequently used as energy storage systems in residential and industrial applications due to their low initial investment cost [16] and low impact on the environment, being easy to recycle [17].

Gel-VRLA technology has a similar structure. The difference from the previous one is that it uses a gelled electrolyte, and the quality of the agent has a great significance on its performance. The gelled sulfuric electrolyte has two main advantages: less or no acid leakage, which implies lower corrosion problems, and flexibility in installation, where the orientation does not affect the performance of the battery [14]. The functionality of Gel-VRLA batteries is governed by their high efficiency, between 90% and 99%, more than 4000 cycles at 40% DoD, long lifetime, with an average of 10 years, good charge stability, and low self-discharge rate [11].

The main disadvantage of VRLA technologies is bad recharge phenomena. These batteries are sensitive to interruptions that can occur during the charging process, especially between 90% and 100%, causing an increase of internal resistance and, in the long term, the reduction of the capacity. The same side effect has the battery remaining in a low discharged state for a long time, resulting also in the loss of active material and a low current value in the initiation of the discharge process [18].

To enhance the performance of VRLA batteries, several electrolyte additives were investigated in recent studies. Inorganic salts and acids added to the electrolyte of LABs could increase the electrochemical performance of the batteries. Boric acid, sodium silicate, citric acid, or phosphoric acid added in specific concentrations in the electrolyte can improve the long-term cycling performance, enhance the rate capability, prevent the corrosion of electrodes, and even inhibit the hydrogen and oxygen reactions [19].

Due to their characteristics, LAB batteries are still the choice in transport and storage of renewable energy, primarily photovoltaic energy. They are also used in the automotive industry, telecommunications, and as a backup source for emergency services.

2.3 Lead-carbon batteries

LCB batteries have significant technical improvements over traditional lead-acid batteries due to the incorporation of carbon into the negative plate, which increases the conductivity, porosity, and inhibits the sulfation of plumb [18]. Having the same schematic layout as the lead-acid batteries (anode, cathode, and electrolyte), the major difference is that there are several carbon-based materials added to negative active materials, which are commonly used in commercialized batteries, to enhance the electrochemical performance: acetylene black, carbon nanofibers, biomass carbon, and graphene [20, 21].

These batteries exceed 7000 cycles during their prolonged lifetime of approximately 20 years at 70% DoD [10] with low maintenance services [22]. The efficiency surpasses 90% [11] and the self-discharge rate is lower than 7% [23]. An important advantage is that 98% of the materials of LCB batteries can be recycled, contributing to reduced lead waste. Most industrialized countries have regulations and programs that recover, recycle, and manufacture new batteries from recycled materials [21].

The main disadvantages include carbon flotation and penetration, causing an increased water leakage during operation [24], low performance at negative temperatures, acid stratification [22], and higher initial investment costs [23].

LCB batteries are usually used in hybrid vehicles, uninterrupted power supply systems [23], golf cart systems, solar street lighting [22], and storage systems for residential roof-top photovoltaic systems [24].

2.4 Vanadium redox flow batteries

VRFB batteries represent a promising solution for reducing grid instability by storing the energy produced from renewable sources [25]. It consists of an anode, a cathode, and two electrolyte solutions from the same species to reduce the risk of contamination, two electrodes, which react with electrolytes to produce electrons, and an ion-selective membrane to separate the electrodes and facilitate ionic conduction. The operating principle involves storing the energy in the redox species, dissolved in the electrolyte. Thus, the battery's ability to store energy depends on the concentration and volume of electrolyte, while the power output is dependent on the number of cell stacks [26], allowing to select the optimal design and capacity to meet the needs of each application.

The main characteristics of VRFB are high energy efficiency, between 75% and 91% [26], negligible self-discharge rate (less than 1%), long durability, and reliability with over 10.000 life cycles [13]. The disadvantages include low power density due to low operating voltage and current density [27], moderate maintenance costs [25], and the high impact on the environment. Vanadium-based catalysts are highly toxic for humans, animals, and plant life due to their heavy metal content. The used VRFB batteries can be recycled. The extraction of vanadium is a process that uses high temperatures and implies toxic emissions [28].

Due to their deep charge and discharge process and many life cycles, VRFB batteries can be used in industrial and residential applications as energy storage systems integrated in photovoltaic systems with installed power ranging from kW scale to MW.

2.5 Aqueous ion hybrid battery

AIHB batteries have a similar structure to Li-ion batteries. It mainly consists of thick positive and negative electrodes, which are made from similar materials. The nonflammable aqueous electrolyte is based on sodium, reducing the energy density and the involved costs [29]. Other characteristics include high efficiency, over 90% and approximately 3000 life cycles at a 70% DoD, which implies a lifetime of around 9 years [11]. The self-discharge process is caused by internal chemical reactions and depends on the purity of the active material and/or electrolyte, electrodes, conductors, binders, or separators. The self-discharge rate is considered between 6% and 9% for AIHB batteries [30].

An important advantage is that AIHB batteries have a low impact on the environment, being considered a good alternative to Li-ion and lead-acid batteries. This compensates for low energy density, which implies an increase in battery mass to obtain a specific capacity [29]. To improve the conductivity, several materials were studied: MoS_2, CuS, ZnS, and other graphene-supported metal sulfides, which theoretically offer high capacity [31].

Despite being in the research-development stage [11], AIHB batteries have been commercialized as an environmentally friendly battery storage system and used primarily as small and medium storage systems in residential photovoltaic and island micro installation applications [29].

2.6 Sodium-sulfur battery

The fundamental structure of a sodium-sulfur battery consists of a molten sulfur positive electrode and a negative molten sodium electrode, separated by a beta alumina ceramic electrolyte, which permits the circulation of sodium ions. To facilitate the movement of ions through the electrolyte, the battery must function at a high temperature, at least 300° C, to ensure good ionic conductivity and maintain both electrodes in liquid state [32]. The particularity is that the cells of the battery are sealed in a thermal enclosure. Due to its structure, this technology is characterized by high power operation (150–230 W/kg [13]) and high energy density.

The main characteristics include a round-trip efficiency of 80% between 5000 and 6000 cycles at 60% DoD, which means approximately 15 years of lifetime. A significant advantage is that the NaS battery has no self-discharge rate [11], has a lower impact on the environment than the Li-ion battery, and is nontoxic [33].

NaS batteries are used as an energy storage system in stationary applications, including residential photovoltaic systems, electric vehicles, the aerospace industry, and forklifts [13]. The developments made in the last decade to improve the functional conditions and overall performances of NaS batteries, enable the integration of this technology as utility-scale storage systems, shifting the wind electricity from off-peak to on-peak [32].

2.7 Proton-exchange membrane fuel cells

The fuel cell structure is similar to the other batteries. It consists of an anode (negative electrode), where the fuel (hydrogen) is fed in, and a cathode (positive electrode), where the air is introduced. The catalyst separates hydrogen into protons and electrons in a polymer electrolyte membrane. The electrons are guided to an external circuit, while the protons reunite with the oxygen in the cathode and produce water [33].

The PEMFC represents a promising electrical energy storage system due to its high power density, high energy conversion efficiency, compact design, flexibility in input fuel, fast startup, lightweight, and low cost [34, 35].

Even though they are still in the research-development stage, PEMFCs are used in both civil and military applications. Due to great stealth capability, low infrared signatures, and noiselessness, PEMFCs are used in submarines as auxiliary power units [35]. Their low environmental impact, feasibility, and performance recommend PEMFC as energy storage systems for off-grid [36], on-grid [37] photovoltaic systems, and electric vehicles [38].

The main technical characteristics are high energy density, with 1.000 Wh/kg and a high number of life cycles, which can exceed 90.000 at a DoD of 80%, depending on operational conditions [39]. Despite recent progress to improve PEMFC performance by replacing the traditional electrolyte membrane with composite ones, there are still challenges regarding the balance of properties to achieve optimal functionality [40].

3. Comparative technical analysis of BESS technologies

This section contains a comparative technical analysis between the analyzed energy storage systems, described in the previous section, based on efficiency [%],

	Li-ion	LAB	LCB	VRFB	AIHB	NaS	PEMFC
Efficiency [%]	90–99 [11]	80–99 [11]	90–99 [11]	75–91 [26]	90 [11]	80 [32]	49.2 [34]
Self-discharge rate [%]	8–31% [11]	2–5 [13]	7% [23]	0.2% [13]	6–9% [30]	None [13]	not applicable [38]
DoD [%]	95–80 [9]	30–50 [11]	70 [21]	not affected	70%	50–60 [32]	80% [39]
Cycles [no.]	6.000 [11]	500–4.000 [11]	7.000 [21]	10.000 [13]	3.000 [11]	5.000–6.000 [13]	6.000–90.000 [39]
Lifetime [years]	5–22 [9]	5–10 [11]	~20 [21]	>20 [28]	9 [11]	15 [32]	10 [34]
Environmental impact	low [11]	low, easy to recycle [17]	very low, 98% recycled [21]	high, hazardous waste [28]	low [11]	low [32]	none [35]
Maintenance	low [11]	low [11]	low [22]	moderate [25]	low [30]	low [33]	moderate [34]

Table 1.
Technical performance of analyzed storage systems.

number of life cycles [no.], depth of discharge [%], self-discharge rate [%], lifetime [years], environmental impact [rank], and need of maintenance during lifetime [rank]. These technical characteristics of the analyzed BESS are presented in **Table 1**.

As can be observed, the Li-ion and LCB batteries have the highest efficiency, which can reach 99%, depending on operational conditions. PEMFC has the lowest value, only 49.2%, compensated by the highest number of life cycles that can reach 90.000. VRFB, LCB, and Li-ion have moderate life cycles, between 6.000 and 10.000.

Considering the efforts made to reduce the carbon footprint on the environment, a significant characteristic of BESS is the environmental impact. PEMFC has no carbon footprint, being the most environmentally friendly BESS, followed by AIHB and NaS. Despite their high percentage of metal, LCB, Li-ion, and LAB are considered eco-friendly due to their recycling capabilities. On the other hand, VRFB are hard to recycle due to the toxicity of the contained materials, having a high impact on the environment.

The most used BESSs in residential photovoltaic applications are Li-ion, LAB, and LCB batteries, due to their technical performances and feasibility. The developments made in recent years increased the interest in AIHB and PEMFC for integration in residential rooftop photovoltaic systems, being considered environmentally friendly solutions that can have a great impact in reducing the carbon footprint in the long term.

4. Feasibility analysis

This chapter was performed to evaluate the feasibility of six BESSs with different capacities and energy storage technologies integrated with a PV system for a residence connected to the utility grid, located in Cluj-Napoca, Romania.

4.1 PV system combined with BESSs simulation

The PVSOL premium 2024 software has been employed for modeling and simulating the electric energy produced by the PV installation. A consumption profile based on hourly real measured values has been generated in the simulation software. The yearly electricity consumed at the examined residence on an hourly basis is presented in **Figure 2**. The dwelling annual electricity consumption is 19,847 kWh. In the next step, grid-integrated PV systems with BESSs have been designed. Shade is an important factor that influences the performance of the PV system. Shading simulations have been carried out to achieve the green electricity produced by the PV installations. The results have been used to evaluate the feasibility of the renewable energy distributed energy generation systems combined with BESSs.

Figure 3 shows the possible one-way or two-way flows of power for the examined renewable energy distributed energy generation systems combined with analyzed energy storage systems. The electric energy produced by the grid-connected solar PV system may be consumed in the residence, stored in an energy storage system, or fed into the power grid.

A PV cell represents a semiconductor device used for converting sunlight into electric energy. PV modules contain cells, usually connected in series. PV modules are wired together in series and in parallel for obtaining PV arrays [41].

PV panels are mainly categorized into two groups: crystalline silicon and thin film [42–45]. Crystalline silicon photovoltaics account for more than 85% of the market [42]. Monocrystalline silicon and polycrystalline silicon are the types of solar energy receptors from this dominant technology. Amorphous silicon, copper indium gallium selenide, and cadmium telluride represent the three main thin-film solar cell technologies.

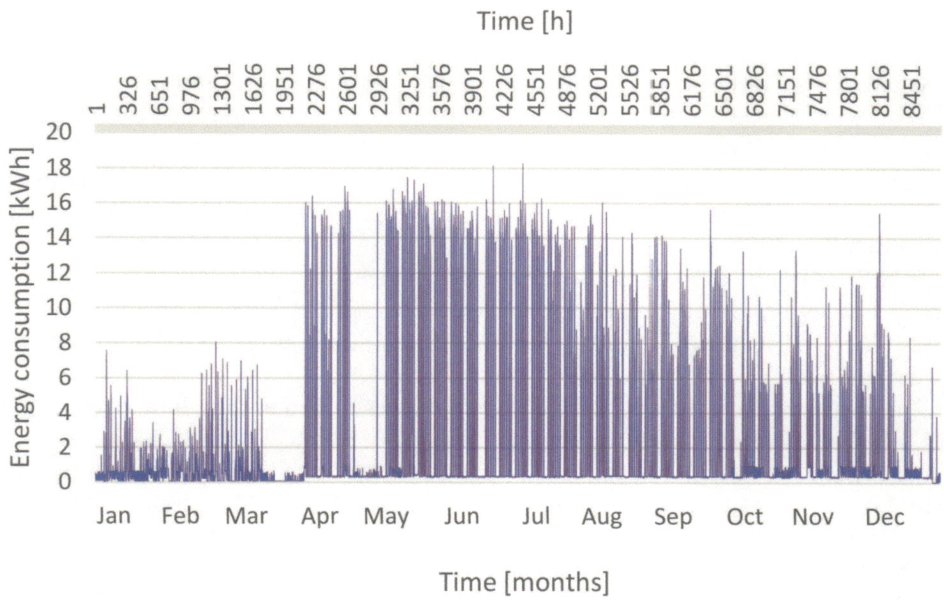

Figure 2.
Yearly electricity consumption on an hourly basis.

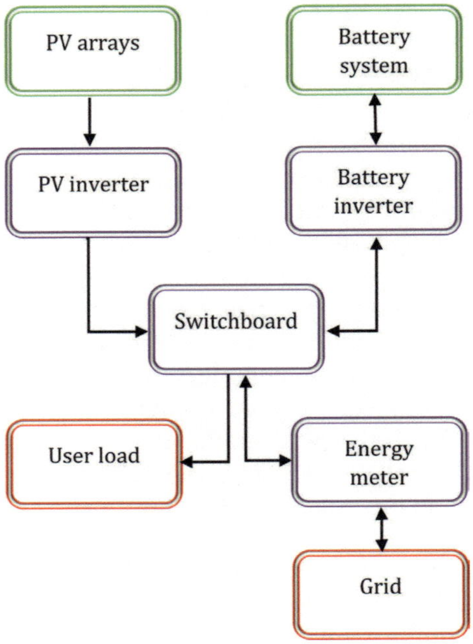

Figure 3.
Flow of power flow for the grid-integrated PV systems with BESS [46, 47].

The output power of the PV array (P_{PV}) mainly depends on rated PV power, geographical location, and meteorological conditions as expressed in the following equation [48, 49]:

$$P_{PV} = ROP_{PV} \cdot DF \cdot \frac{G_T}{G_{T,\,STC}} \cdot [1 + \beta_P \cdot \left(T_C - T_{C,\,STC}\right) \qquad (1)$$

where ROP_{PV} is the array's rated output power at standard test condition (STC), DF represents the derating factor, G_T is the solar irradiation incident on the PV array's plane, $G_{T,STC}$ is the solar irradiation incident on the PV array's plane, $G_{T,STC}$ represents the solar irradiation at STC conditions (1 kW/m^2), β_P represents the power's temperature coefficient, T_C is the PV cell temperature, $T_{C,STC}$ represents the PV cell temperature under STC (25°C).

In this research, monocrystalline panels have been chosen due to their efficiency, longevity, maintenance cost, space requirement efficiency, and eco-friendliness [50, 51]. The most important technical specifications of the PV modules are shown in **Table 2**.

The total installation costs of the PV system investigated in this chapter are $30,000 USD. The used currency exchange rate is 4.5582 RON for $1 USD as of August 2024. The considered prices have been decided by taking into account the quoted prices collected at the beginning of 2024 from 10 solar PV installation firms from Romania. The annual operation and maintenance costs represent one percent of the total investment [48] including also the replacement cost [52]. The renewable energy distributed energy generation systems consist of 40 monocrystalline modules, oriented South, with a tilt angle of 33°.

BESSs offer a range of benefits such as diminishing energy costs, increasing self-consumption, reliability, and reducing the carbon footprint. The electrical energy

Characteristics	**Value**
Maximum power (P_{max}) [W]	500
Voltage at Pmax [V]	45
Current at Pmax [A]	11.12
Open vircuit voltage (V_{oc}) [V]	53.7
Short circuit current (I_{sc}) [A]	11.77
Panel efficiency [%]	21.2
Operating temperature range	−40 °C ~ +85°C
Temperature coefficient of P_{max} [%/°C]	−0.35
Temperature coefficient of V_{oc} [%/°C]	−0.27
Temperature coefficient of I_{sc} [%/°C]	0.05
Degradation per year [%]	0.6
Panel dimension (length × width × height) [mm]	2250x1048x35
Lifetime [years]	25

Table 2.
Main PV modules' technical specifications.

stored in the battery depends on the discharge η_d and charge η_c efficiencies (%). Thus, the stored energy (kWh) in BESS at time *t*, *E(t)*, is defined as [53]:

$$E(t) = E(t-1) - \frac{\Delta T \cdot P_d(t)}{\eta_d} - \Delta T \cdot \eta_c \cdot P_c(t) \tag{2}$$

where *E(t-1)* is the energy stored (kWh) at time interval *t-1*, $P_d(t)$ and $P_c(t)$ are the discharge and charge powers (kW) at time *t* for the period of time ΔT (h) [53].

The state of charge (SOC) of a battery shows the energy available at any time, and it is determined considering the stored energy at time *t*, *E(t)*, and the considered BESS capacity, $BESS_{capacity}$. The *SOC(t)* must always be larger or equal to the minimum SOC, SOC_{min} (%), determined by the depth of discharge specific to each battery technology, and smaller or equal to the maximum SOC, SOC_{max} (%), to avoid overcharging the BESS. These constraints are to protect the battery and prolong the lifetime of BESS. The equations regarding *SOC* are the following [53]:

$$SOC(t) = \frac{E(t)}{BESS_{capacity}} \cdot 100 \tag{3}$$

$$SOC_{min} \leq SOC(t) \leq SOC_{max} \tag{4}$$

The battery is equipped with a converter, which connects the BESS with a home energy management system. This controls the battery that distributes energy to home appliances. The charged/discharged electricity from the battery at time *t*, $P_{BESS}(t)$, depends on the power convertor efficiency, η_{conv} (%), and can be calculated with the following equation [53]:

$$P_{BESS}(t) = P_d(t) \cdot \eta_{conv} + \frac{P_c(t)}{\eta_{conv}} \cdot 100 \quad (5)$$

In this study, the performance of the PV installation has been assessed together with six different capacities of energy storage systems. **Table 3** shows the main features of the examined Li-ion batteries. The investigated BESSs are lithium-iron phosphate ($LiFePO_4$). The depth of discharge for each version is 100%.

In this research, the costs for the BESS I, II, and III are $5500 USD, $7850 USD, and $11,000 USD, respectively. The warranty period for the examined Li-ion batteries is 10 years. The service life for BESS I, II, and III is 15 years, 18 years, and 21 years. Therefore, it is necessary to substitute the BESSs after the service life. The BESS cost is forecasted to decline at a rate of 5% annually of the present price [54]. The annual operation and maintenance costs for BESS I, II, and III are 0.5% of the initial investment [55].

Table 4 shows the main features of the examined lead-acid batteries. **Table 4** are presents the main features of the examined lead-acid batteries.

In this research, the costs for the BESS IV, V, and VI are $2484 USD, $3864 USD and $4968 USD, respectively. The service life for the BESS IV is four years and for BESS V and VI is five years. As in the case of Li-ion batteries, it is necessary to substitute the BESSs after the service life. Also, the BESS cost is forecasted to decline at a rate of 5% annually of the present price [54]. The annual operation and maintenance costs for the lead-acid batteries are 1%of the initial investment [55].

The schematic circuit diagram of the examined PV system is shown in **Figure 4**. The distributed energy generation system with a rated power of 20 kW includes 40 PV modules with a capacity of 500 W.

The PV panels have been arranged in four parallel strings, each containing 10 modules, as illustrated in **Figure 4**.

Characteristics	BESS version		
	I	II	III
Usable energy [kWh]	10	15	20.7
Maximum SOC [%]	100	100	100
Nominal voltage [V]	51.2	51.2	25.6
Battery voltage [V]	102.4	153.6	76.8

Table 3.
Main specifications of Li-ion batteries.

Characteristics	BESS version		
	IV	V	VI
Usable energy [kWh]	10.8	16.8	21.6
Maximum SOC [%]	90	90	90
Nominal voltage [V]	6	6	6
Battery voltage [V]	24	48	48

Table 4.
Main specifications of lead-acid batteries.

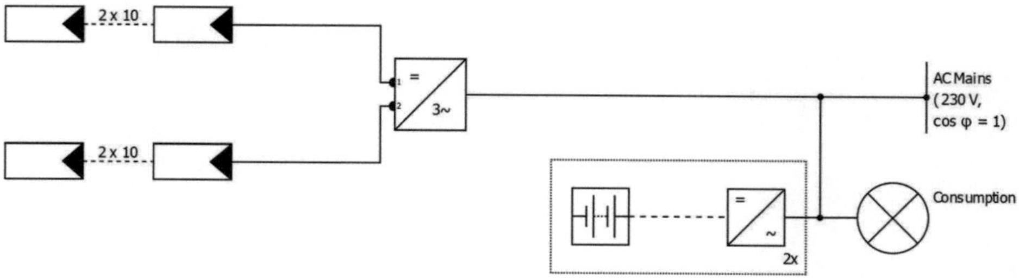

Figure 4.
Schematic circuit diagram of the PV system combined with BESS.

4.2 Economic assessment

The net present value (NPV) represents the method that has been employed to evaluate the economic feasibility of the renewable energy distributed energy generation systems combined with BESSs because it is primarily used in the assessment of investment in projects [56]. This method calculates the present value of the cash flows from an investment in a project. NPV can be determined as follows:

$$NPV = \sum_{i=1}^{n} \frac{CF_i}{(1+r)^i} - C_0 \qquad (6)$$

Where, CF_i is the cash flow from year i; r represents the discount rate; C_0 is the initial investment cost; n represents the lifetime of the investment.

The yearly cashflows include the value of electric energy injected into the grid, the electricity bill savings by installing BESS integrated with PV installation, and costs associated with operation and maintenance. A discount rate of 8% has been considered in evaluating the investment projects. The prosumer pays about \$0.2852 USD for every kWh of electric energy used from the grid, while the price for a kWh of electricity injected into the network is approximately \$0.1464 USD.

Investment in a project is feasible if its NPV is greater than zero. Otherwise, if the NPV of the investment in a project is smaller than zero, then it is not viable.

4.3 Results

The economic feasibility represents one of the most important steps to be taken for the success of families' investment in installing a grid-integrated PV system with BESS.

Figure 5 illustrates the coverage of electricity consumption per month when the PV system is combined with the three Li-ion batteries. As can be seen in **Figure 5**, the electric energy produced by renewable energy distributed energy generation systems combined with BESSs is not sufficient to cover the consumption throughout the year. August is the month in which the photovoltaic system together with BESS I covers most of the consumption in a proportion of almost 70%, while in April only 50% of the consumption is covered by green energy. It can be observed that BESSs determine the way electricity consumption is covered. The lower their usable energy is, the higher the amount of electric energy bought from the grid.

Figure 6 presents the coverage of monthly electricity consumption when the PV system is combined with the three lead-acid batteries. It can be observed that the

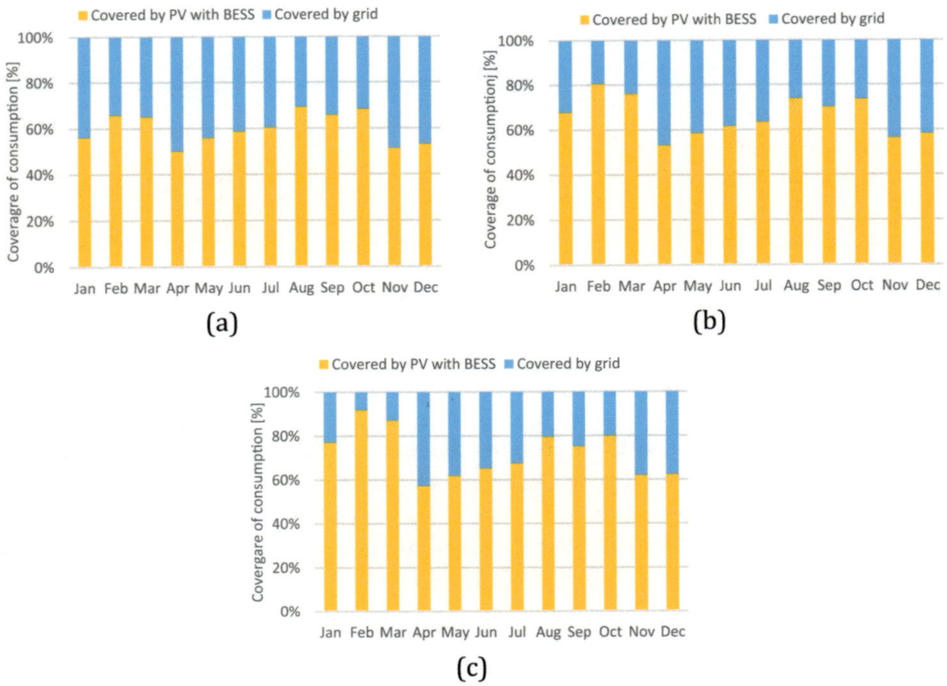

Figure 5.
Coverage of energy consumption for (a) BESS I, (b) BESS II, (c) BESS III.

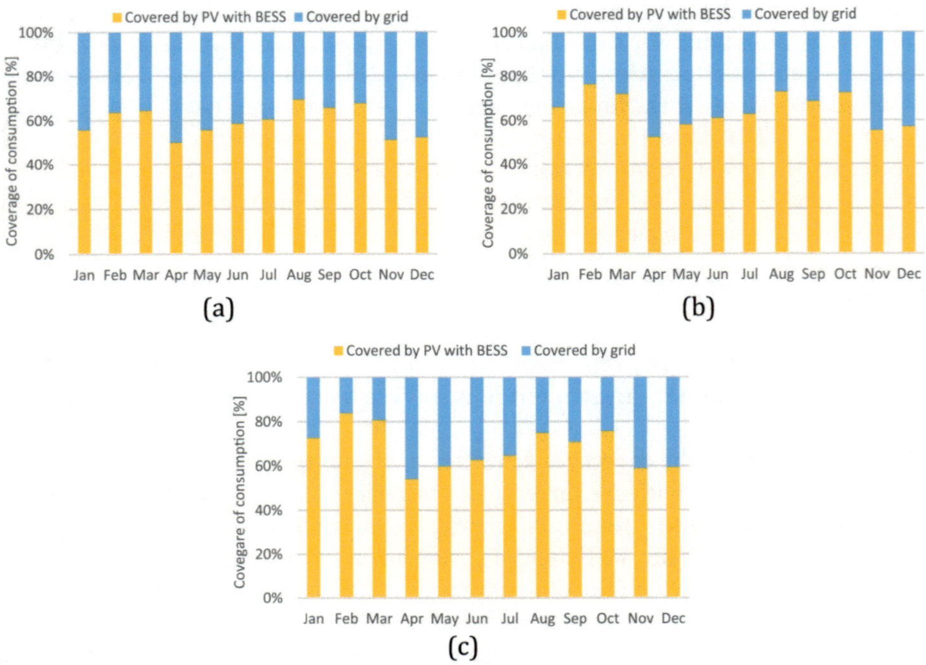

Figure 6.
Coverage of energy consumption for (a) BESS IV, (b) BESS V, (c) BESS VI.

higher the lead-acid batteries' usable energy is, the smaller the amount of electricity bought from the grid. Also, it is observed that lead-acid batteries cover less of the residence's consumption compared to Li-ion batteries.

Figure 7 shows how the electricity produced by the PV system combined with the three Li-ion batteries is being used. November is the month when most of the electricity produced by the photovoltaic system together with BESS III (more than 72%) is

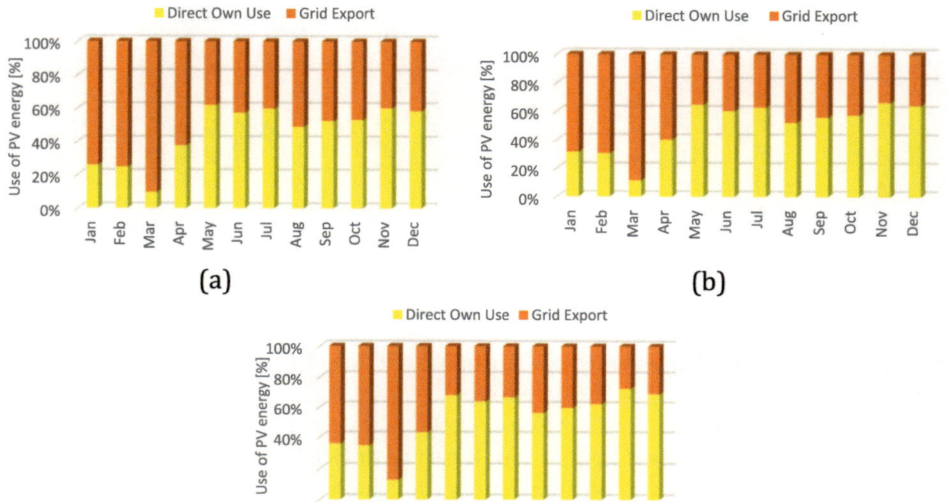

Figure 7.
Use of PV energy for (a) BESS I, (b) BESS II, (c) BESS III.

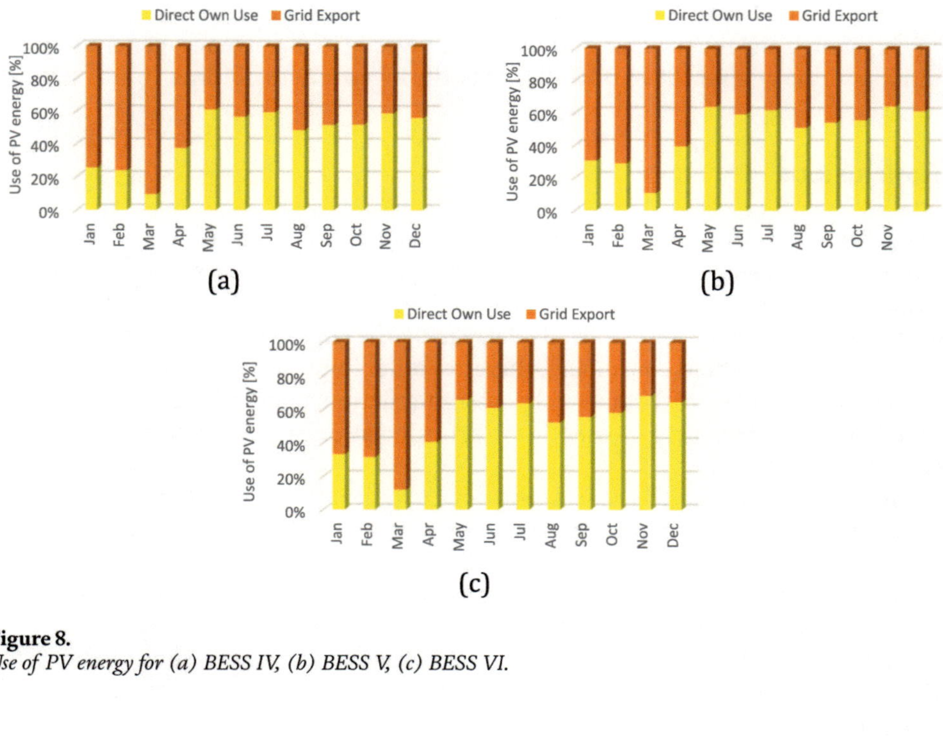

Figure 8.
Use of PV energy for (a) BESS IV, (b) BESS V, (c) BESS VI.

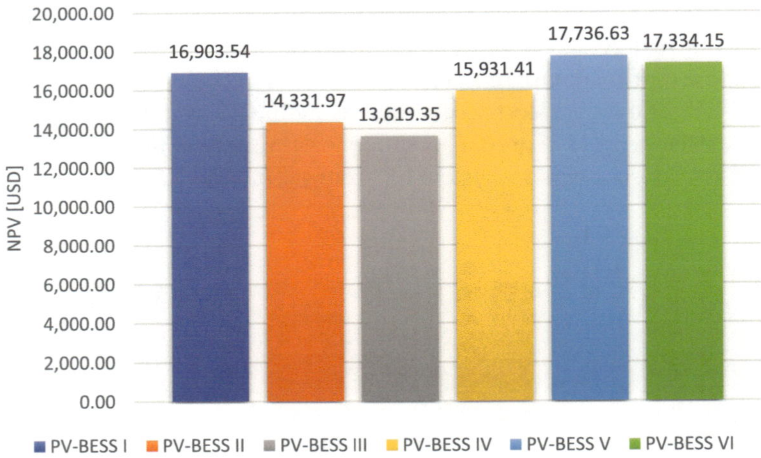

Figure 9.
NPV for PV system combined with BESSs.

directly consumed by the residence, while in March only 13% of the green electric energy is directly consumed in the dwelling.

Figure 8 illustrates how the electricity produced by the PV system combined with the three lead-acid batteries is being used. It can be noticed that the higher the usable energy is, the greater the share of green energy produced is consumed directly in the residence. It is also seen that with lead-acid batteries, the proportion of green energy produced and consumed directly in the residence is lower than with Li-ion batteries.

Figure 9 presents the NPV for the renewable energy distributed energy generation systems combined with BESSs. It can be observed that a grid-connected residential PV system coupled to all BESSs is economically feasible, but BESS V integrated with the PV system is the most viable option for the investigated residence, generating an additional value of almost $18,000 USD over the investment's lifetime.

5. Conclusions

The integration of BESS in residential photovoltaic systems represents a feasible solution for increasing the consumption of energy from renewable sources, storing the energy surplus, and using it when needed. For the household consumer, this translates into reduced energy bills. By decreasing the energy quantity injected into the grid by the household consumers, there are other important positive effects, such as reduced grid disturbances and power fluctuations, minimized peak energy demand, and reduced energy losses due to transportation and distribution of electrical energy.

This chapter investigated the feasibility of integrating six BESSs in a grid-connected solar PV installation for a residence situated in the north-western part of Romania. The obtained results reveal that all investigated BESSs integrated with the PV system are viable, but the PV installation combined with the 16.8 kWh lead-acid battery is the most feasible variant for the examined dwelling, generating an additional value of nearly $18,000 USD over the investment project's lifetime. Although with Li-ion batteries, the proportion of green energy produced and consumed directly in the residence is higher than with lead-acid batteries and, at the same time, investigated Li-ion batteries cover more of the home's consumption compared to lead-acid batteries, the most viable option is a lead-acid battery, mainly because Li-ion batteries require large capital expenditures.

The attained results may assist Romanian stakeholders in acknowledging the economic feasibility of grid-connected PV systems combined with different BESSs technologies. Also, the findings presented in this chapter may support families that intend to invest in a BESS integrated with a PV installation. This chapter assessed the feasibility of a PV system combined with Li-ion and lead-acid batteries. Future research may focus on investigating other BESSs technologies.

Acknowledgements

This research received funding for open access publishing from the authors' employer, the Technical University of Cluj-Napoca.

Author details

Ciprian Cristea*, Maria Cristea, Radu-Adrian Tîrnovan and Florica Șerban
Faculty of Electrical Engineering, Technical University of Cluj-Napoca, Cluj-Napoca, Romania

*Address all correspondence to: ciprian.cristea@emd.utcluj.ro

References

[1] Soto EA, Bosman LB, Wollega E, Leon-Salas WD. Analysis of grid disturbances caused by massive integration of utility level solar power systems. Eng. 2022;**3**(2):236-253. DOI: 10.3390/eng3020018

[2] Nikolaidis P, Poullikkas A. Cost metrics of electrical energy storage technologies in potential power system operations. Sustainable Energy Technologies and Assessments. 2018;**25**:43-59. DOI: 10.1016/j.seta.2017.12.001

[3] Das CK, Bass O, Kothapalli G, Mahmoud TS, Habibi D. Overview of energy storage systems in distribution networks: Placement, sizing, operation, and power quality. Renewable and Sustainable Energy Reviews. 2018;**91**:1205-1230. DOI: 10.1016/j.rser.2018.03.068

[4] Ashfaq S, El Myasse I, Musleh AS, Zhang D, Dong ZY. Least cost analysis of bulk energy storage for deep decarbonized power system with increased share of renewable energy. Electric Power Systems Research. 2023;**220**:109375. DOI: 10.1016/j.epsr.2023.109375

[5] Ashfaq S, El Myasse I, Zhang D, Musleh AS, Liu B, Telba AA, et al. Comparing the role of long duration energy storage technologies for zero-carbon electricity systems. IEEE Access. 2024;**12**:73169-73186. DOI: 10.1109/ACCESS.2024.3397918

[6] Mousavi GSM, Faraji F, Majazi A, Al-Haddad K. A comprehensive review of flywheel energy storage system technology. Renewable and Sustainable Energy Reviews. 2017;**67**:477-490. DOI: 10.1016/j.rser.2016.09.060

[7] Wang SY, Lu JL. Optimal sizing of the CAES system in a power system with high wind power penetration. Electrical Power and Energy Systems. 2012;**37**:117-125. DOI: 10.1016/j.ijepes.2011.12.015

[8] Fratean A, Dobra P. Technical and economic viability of greenfield large scale photovoltaic plants in Romania. Sustainable Energy Technologies and Assessments. 2022;**53**. DOI: 10.1016/j.seta.2022.102486

[9] Vonsien S, Madlener R. Li-ion battery storage in private households with PV systems: Analyzing the economic impacts of battery aging and pooling. Journal of Energy Storage. 2020;**29**. DOI: 10.1016/j.est.2020.101407

[10] Marashli A, Al-Kassab AI, Gab-Allah DM, Shalby M, Salah A. Numerical life cycle assessment of lithium ion battery, Li-NMC type, integrated with PV system. Results in Engineering. 2024;**23**:102489. DOI: 10.1016/j.rineng.2024.102489

[11] Cristea M, Tîrnovan R, Cristea C, Fagarasan C. Levelized cost of storage (LCOS) analysis of BESSs in Romania. Sustainable Energy Technologies and Assessments. 2022;**53**. DOI: 10.1016/j.seta.2022.102633

[12] Martinez-Bolanos JR, Udaeta Morales ME, Veiga Gimenes AL, Oliveira da Silva V. Economic feasibility of battery energy storage systems for replacing peak power plants for commercial consumers under energy time of use tariffs. Journal of Energy Storage. 2020;**29**. DOI: 10.1016/j.est.2020.101373

[13] Salkuti SR. Comparative analysis of electrochemical energy

storage technologies for smart grid. TELKOMNIKA Telecommunication, Computing, Electronics and Control. 2020;**18**(4):2118-2124. DOI: 10.12928/TELKOMNIKA.v18i4.14039

[14] Tantichanakul T, Chailapakul O, Tantavichet N. Influence of fumed silica and additives on the gel formation and performance of gel valve-regulated lead-acid batteries. Journal of Industrial and Engineering Chemistry. 2023;**19**:2085-2091. DOI: 10.1016/j.jiec.2013.03.024

[15] Kumar V, Kameswara Rao PV, Rawal A. Amplification of electrolyte uptake in the absorptive glass mat (AGM) separator for valve regulated lead acid (VRLA) batteries. Journal of Power Source. 2017;**341**:19-26. DOI: 10.1016/j.jpowsour.2016.11.027

[16] Numbi BP, Malinga SJ. Optimal energy cost and economic analysis of a residential grid-interactive solar PV system- case of eThekwini municipality in South Africa. Applied Energy. 2017;**186**:28-45. DOI: 10.1016/j.apenergy.2016.10.048

[17] Tsafack P, Fru SE, Nghemachi AV, Tanyi E. Impact of high constant charging current rates on the charge/discharge efficiency in lead acid batteries, for residential photovoltaic system applications. Journal of Energy Storage. 2023;**63**. DOI: 10.1016/j.est.2023.107013

[18] Wieczorek M, Wodyk S, Poliszkiewicz R, Witaszek P. The influence of current in off-grid PV systems on lead–acid battery lifetime and hybridization with LFP battery as solution. Energy Reports. 2023;**9**:766-773. DOI: 10.1016/j.egyr.2023.05.213

[19] Mosallanejad B, Akrami M. Recent advances on electrolyte additives used in lead-acid batteries to enhance their electrochemical performances. Journal of Energy Storage. 2024;**97**. DOI: 10.1016/j.est.2024.112738

[20] He P, Tu J, Yang Y, Huang H, Chen B, Gao C, et al. Hierarchical tubular porous carbon derived from mulberry branches for long-life lead-carbon battery. Journal of Energy Storage. 2023;**64**. DOI: 10.1016/j.est.2023.107162

[21] Wang Z, Tuo X, Zhou J, Xiao G. Performance study of large capacity industrial lead-carbon battery for energy storage. Journal of Energy Storage. 2022;**55**. DOI: 10.1016/j.est.2022.105398

[22] Vangapally N, Penki TR, Elias Y, Muduli S, Maddukuri S, Luski S, et al. Lead-acid batteries and lead–carbon hybrid systems: A review. Journal of Power Sources. 2023;**579**. DOI: 10.1016/j.jpowsour.2023.233312

[23] Arun S, Mithin Kumar S, Uday Venkat Kiran K, Sundar M. Effects of carbon surface area and morphology on performance of stationary lead acid battery. Journal of Energy Storage. 2020;**32**. DOI: 10.1016/j.est.2020.101763

[24] He P, He Y, Yang Y, Huang H, Guo Z. In-situ carbon encapsulated Pb/PbO nanoparticles derived from spent lead paste for lead-carbon battery. Journal of Power Sources. 2024;**614**. DOI: 10.1016/j.jpowsour.2024.234989

[25] Dai G, Huang Y, Chu F, Jin C, Liu H. Analysis of the effect of thermal treatment and catalyst introduction on electrode performance in vanadium redox flow battery. Heliyon. 2024;**10**. DOI: 10.1016/j.heliyon.2024.e33561

[26] Brahma K, Nayak R, Verma SK, Sonika. Recent advances in development and application of polymer

nanocomposite ion exchange membrane for high performance vanadium redox flow battery. Journal of Energy Storage. 2024;**97**. DOI: 10.1016/j.est.2024.112850

[27] Wang J, Mu A, Yang B, Wang Y, Wang W. Numerical simulation of all-vanadium redox flow battery performance optimization based on flow channel cross-sectional shape design. Journal of Energy Storage. 2024;**93**. DOI: 10.1016/j.est.2024.112409

[28] Beriwal N, Sharma L, Verma A. Powering a vanadium redox flow battery using spent vanadium catalyst: Extraction of direct-use V (IV)/V(III) vanadium precursors. Journal of Cleaner Production. 2023;**429**. DOI: 10.1016/j.jclepro.2023.139568

[29] Peters JF, Weil M. Aqueous hybrid ion batteries e an environmentally friendly alternative for stationary energy storage? Journal of Power Sources. 2017;**364**:258-265. DOI: 10.1016/j.jpowsour.2017.08.041

[30] Konarov A, Gosselink D, Zhang Y, Tian Y, Askhatova D, Chen P. Self-discharge of rechargeable hybrid aqueous battery. ECS Electrochemistry Letters. 2015;**4**:A151-A154. DOI: 10.1149/2.0111512eel

[31] Tan M, Qin J, Wang Y, Zhang F, Zhao X, Lei X. CuS loaded on reduced graphene oxide prepared by ball milling method as cathode material for high- power aqueous Cu-Al hybrid-ion batteries. Electrochimica Acta. 2024;**476**:143734. DOI: 10.1016/j.electacta.2023.143734

[32] Rodrigues EMG, Osorio GJ, Godina R, Bizuayehu AW, Lujano-Rojas JM, Matias JCO, et al. Modelling and sizing of NaS (sodium sulfur) battery energy storage system for extending wind power performance in Crete Island. Energy. 2015;**90**:1606-1617. DOI: 10.1016/j.energy.2015.06.116

[33] Landi D, Marconi M, Pietroni G. Comparative life cycle assessment of two different battery technologies: Lithium iron phosphate and sodium-sulfur. Procedia CIRP. 2022;**105**:482-488. DOI: 10.1016/j.procir.2022.02.080

[34] Nasser M, Hassan H. Assessment of standalone streetlighting energy storage systems based on hydrogen of hybrid PV/electrolyzer/fuel cell/desalination and PV/batteries. Journal of Energy Storage. 2023;**63**. DOI: 10.1016/j.est.2023.106985

[35] Meng K, Zhou H, Chen B, Tu Z. Dynamic current cycles effect on the degradation characteristic of a H_2/O_2 proton exchange membrane fuel cell. Energy. 2021;**224**. DOI: 10.1016/j.energy.2021.120168

[36] Soyturk G, Kizilkan O, Ezan MA, Colpan CO. Design, modeling, and analysis of a PV/T and PEM fuel cell based hybrid energy system for an off-grid house. International Journal of Hydrogen Energy. 2024;**67**:1181-1193. DOI: 10.1016/j.ijhydene.2023.11.291

[37] Haddad A, Jaber H, Khaled M, Al Afif R, Ramadan M. An investigation on coupling fuel cell, wind turbine and PV as green to green system. International Journal of Hydrogen Energy. 2024;**52**:923-932. DOI: 10.1016/j.ijhydene.2022.06.107

[38] Cano ZP, Banham D, Ye S, Hintennach A, Lu J, Fowler M, et al. Batteries and fuel cells for emerging electric vehicle markets, nature. Energy. 2018;**3**:279-289. DOI: 10.1038/s41560-018-0108-1

[39] The Comparison of Different Technologies for On-board Power Supply of Cruise Ships, Nautical Integrated

Hybrid Energy System for Long-haul Cruise Ships Project 2020. Available from: https://nautilus-project.eu/news/the-comparison-of-different-technologies-for-on-board-power-supply-of-cruise-ships [Accessed: July 27, 2024]

[40] Qasem NAA. A recent overview of proton exchange membrane fuel cells: Fundamentals, applications, and advances. Applied Thermal Engineering. 2024;**252**:123746. DOI: 10.1016/j.applthermaleng.2024.123746

[41] Tian H, Mancilla-David F, Ellis K, Muljadi E, Jenkins P. A cell-to-module-to-array detailed model for photovoltaic panels. Solar Energy. 2012;**86**(9):2695-2706. DOI: 10.1016/j.solener.2012.06.004

[42] Hajialigol P, Moazami A, Aghaei M. Solar energy system integration for energy transition: A short review from technologies and methods to energy management system and challenges. In: Aghaei M, Moazami A, editors. Solar Radiation—Enabling Technologies, Recent Innovations, and Advancements for Energy Transition. London, UK: IntechOpen; 2024. DOI: 10.5772/intechopen.114956

[43] Jasim, Hasan AlTimimi M. Solar Energy. In: Thirumalai J, editor. Quantum Dots—Recent Advances, New Perspectives and Contemporary Applications. London, UK: IntechOpen; 2023. DOI: 10.5772/intechopen.106155

[44] Takele GF. Solar energy conversion efficiency, growth mechanism and design of III–V nanowire-based solar cells: Review. In: Ismail BI, editor. Solar PV Panels—Recent Advances and Future Prospects. London, UK: IntechOpen; 2023. DOI: 10.5772/intechopen.105985.parallel

[45] Ibrahim, Eisa Idris B, Mohamed A, Salah A, et al. Application of solar energy in medical instruments (microscope). In: Elseman AM, editor. Solar Cells—Theory, Materials and Recent Advances. London, UK: IntechOpen; 2021. DOI: 10.5772/intechopen.97177

[46] Angenendt G, Zurmuhlen S, Axelsen H, Sauer DU. Prognosis-based operating strategies for smart homes with power-to-heat applications. Energy Procedia. 2018;**155**:136-148. DOI: 10.1016/j.egypro.2018.11.061

[47] Weniger J, Tjaden T, Quasching V. Sizing of residential PV battery systems. Energy Procedia. 2014;**46**:78-87. DOI: 10.1016/j.egypro.2014.01.160

[48] Özcan Ö, Duman AC, Gönül Ö, Güler Ö. Techno-economic analysis of grid-connected PV and second-life battery systems for net-zero energy houses. Journal of Building Engineering. 2024;**89**:109324. DOI: 10.1016/j.jobe.2024.109324

[49] Al-Masri HMK, Magableh SK, Abuelrub A. Output power computation and sizing of a photovoltaic array by advanced modeling. Sustainable Energy Technologies and Assessments. 2021;**47**:101519. DOI: 10.1016/j.seta.2021.101519

[50] Lim MSW, He D, Tiong JSM, Hanson S, Yang TKG, Tiong TJ, et al. Experimental, economic and life cycle assessments of recycling end-of-life monocrystalline silicon photovoltaic modules. Journal of Cleaner Production. 2022;**340**:130796. DOI: 10.1016/j.jclepro.2022.130796

[51] Zhou X, Yu N, Wu Z, Maleki A. Energy control and design optimization of a hybrid solar-hydrogen energy storage system using various solar panel technologies. Journal of Energy

Storage. 2024;**94**:112389. DOI: 10.1016/j.est.2024.112389

[52] Dehwah AHA, Asif M. Assessment of net energy contribution to buildings by rooftop photovoltaic systems in hot-humid climates. Renewable Energy. 2019;**131**:1288-1299. DOI: 10.1016/j.renene.2018.08.031

[53] Elkazaz M, Sumner M, Naghiyev E, Pholboon S, Davies R, Thomas D. A hierarchical two-stage energy management for a home microgrid using model predictive and real-time controllers. Applied Energy. 2020;**269**:115118. DOI: 10.1016/j.apenergy.2020.115118

[54] Nykvist B, Nilsson M. Rapidly falling costs of battery packs for electric vehicles. Nature Climate Change. 2015;**5**:329-332. DOI: 10.1038/nclimate2564

[55] Wang W, Kang K, Sun G, Xiao L. Configuration optimization of energy storage and economic improvement for household photovoltaic system considering multiple scenarios. Journal of Energy Storage. 2023;**67**:107631. DOI: 10.1016/j.est.2023.107631

[56] Cristea C, Cristea M, Birou I, Tîrnovan RA. Economic assessment of grid-connected residential solar photovoltaic systems introduced under Romania's new regulation. Renewable Energy. 2020;**162**:13-29. DOI: 10.1016/j.renene.2020.07.130

Section 3

Electrochemical and Thermal Performance Analysis of Energy Storage Mediums

Chapter 4

Electrochemical Performance and Design Strategies of MXene-Based Supercapacitors

Gopi R.R., Amritha Kadathamana, Jithin Varghese, Shadai Lugo Loredo, Andrea Cerdán-Pasarán, Soorya Pushpan, J.A. Hernández-Magallanes and Sanal K.C.

Abstract

Supercapacitors, at the forefront of energy storage, bridge the gap between traditional capacitors and batteries, renowned for high power densities, rapid charge/discharge rates, and extended cycle life. MXenes, two-dimensional transition metal carbides, nitrides, and carbonitrides, have emerged as promising supercapacitor electrode materials due to remarkable electrical conductivity and tunable surface properties. This chapter explores MXene-based supercapacitors, detailing synthesis methods, electrochemical performance, and potential applications. MXene-based supercapacitor performance, covering specific capacitance, energy density, power density, and cycling stability, is discussed. The chapter also addresses challenges in MXene-based supercapacitor development, including scalability, cost-effectiveness, and long-term stability, applications in portable electronics, wearables, and electric transportation. Providing a comprehensive overview, this chapter inspires further exploration and utilization of MXene-based supercapacitors for next-gen energy storage solutions.

Keywords: two-dimensional materials, MXene, composite, energy storage devices, super capacitors

1. Introduction

Attaining net-zero carbon emissions is a universal objective shared by nations across the globe. A significant contributor to carbon emissions is vehicular exhaust, which has led to the rise of Electric Vehicles (EVs) as a modern alternative to traditional cars. This shift toward EVs is a promising start in the battle against carbon emissions. However, it's crucial to note that the electricity powering these vehicles often originates from power plants reliant on non-renewable resources, which themselves contribute to carbon emissions and other environmental concerns. Therefore, the shift toward renewable energy sources and the production of more efficient storage devices is essential to achieving the goal of net-zero carbon emissions. In this context, the role of advancements in material science and electrochemical energy storage devices becomes paramount.

Energy storage technologies, such as batteries and supercapacitors, fall into two main categories: electrochemical and electronic devices. Recent research has focused on enhancing the performance and rate capabilities of rechargeable batteries and electrochemical capacitors. Supercapacitors demonstrate substantial potential for future energy storage applications due to their high power density and rapid charge-discharge cycles. Furthermore, emerging technologies like pseudocapacitors and hybrid devices, known as "supercapatteries," offer a promising balance between high energy density and impressive rate capability.

To develop an efficient supercapacitor material with a high surface area, porous structure, good electrical conductivity, and flexibility are needed. It often employs metal compounds, conductive polymers, and carbon-based materials as their constituent ingredients. For example, because of their superior conductivity and great surface area, carbon compounds such as carbon nanotubes and porous active carbon are frequently utilized [1]. The flexibility of conducting polymers like polythiophene, polypyrrole, and polyaniline makes them appealing as well [2]. Recently, a class of materials known as MXene, which includes metal carbides and nitrides, has been the subject of extensive research for a range of energy-related uses. Because of their higher volumetric capacitances, broad surface area, and electrical conductivity, MXenes have been extensively investigated as supercapacitor electrodes. Also, MXene composites are adaptable to many environments and industries, driving ongoing research and development to maximize their material science and technology potential. The good mechanical strength of MXenes can be utilized to make flexible and wearable supercapacitors, which can be incorporated with fabrics for the real-time monitoring of physiological conditions. However, more research must be done on this topic, as various factors are to be considered for wearable devices. Innovation from the lab to the market requires commercialization and industrial applications. This process turns R&D into market-ready solutions that solve real-world problems and meet consumer needs.

This chapter aims to highlight the importance of MXenes in revolutionizing energy storage technologies essential for sustainable development and net-zero carbon emissions. By exploring MXenes' structure, synthesis methods, and unique electrochemical properties—such as high capacitance, rate capability, and cyclic stability—this chapter establishes MXenes as a pivotal material for the next generation of supercapacitors and flexible, wearable devices. Their versatility and mechanical strength make them ideal for integration into scalable energy storage solutions that can support renewable energy sources. Ultimately, this chapter seeks to bridge fundamental knowledge and practical application, underscoring MXenes' potential to drive innovation in sustainable energy storage systems, which are critical to achieving future energy and environmental goals.

2. MXenes: Structure and synthesis

MXenes are two-dimensional materials consisting of transition metal carbides and nitrides that serve as graphene's replacement because of their exceptional qualities, which include excellent conductivity, high tensile strength, hydrophilicity, flexibility, and customizable surface termination [3, 4]. By altering the element, structure, and composition, numerous MXenes have been discovered with various dominating properties. Currently there are 30 different experimentally made stoichiometric MXenes and almost 250+ combinations of MXenes, but still $Ti_3C_2T_x$ is the one studied mostly [5, 6]. To utilize its potential, it is necessary to understand the structure and relative properties of MXenes.

2.1 The crystal structure of MXenes

The chemical formula of MXene is $M_{n+1}X_nT_x$ (n = 1, 2, and 3), where M stands for transition metals, X stands for carbon or nitrogen, which occupy the octahedral sites between the neighboring transition metal sites, and T refers to surface factional groups such as –O, –F, or –OH. Usually, MXene has a hexagonal structure, and the ordering of the layers depends on the n value [7]. Commonly seen four different configurations of MXene are M_2XT_x, $M_3X_2T_x$, $M_4X_3T_x$, and $M_5X_4T_x$.

The MXene is usually derived from the MAX phase, which has a ternary formula of $M_{n+1}AX_n$, self-possessed with two or more layers of transition metals, where A stands for group 13 or 14 elements and X is carbon or nitrogen layers, where n = 1–4 and usually has hexagonal structure ($P6_3/mmc$ symmetry) [9, 10]. **Figure 1** shows the structure of the MAX phase of binary, ternary, and quaternary. The chemical bonding between transition metal and carbides/nitrides (M-X) is an ionic, or covalent bond, whereas the bond between M–A is purely metallic, which is easy to break by etching it with suitable etching.

The MAX phases are widely introduced and investigated by Barum et al. in various compositions that include Ti_2AlC, Ti_3GeC_2, Ti_4AlN_3 [11], which have hexagonal structure ($P6_3/mmc$ symmetry). Later, MAX phase with two different transition

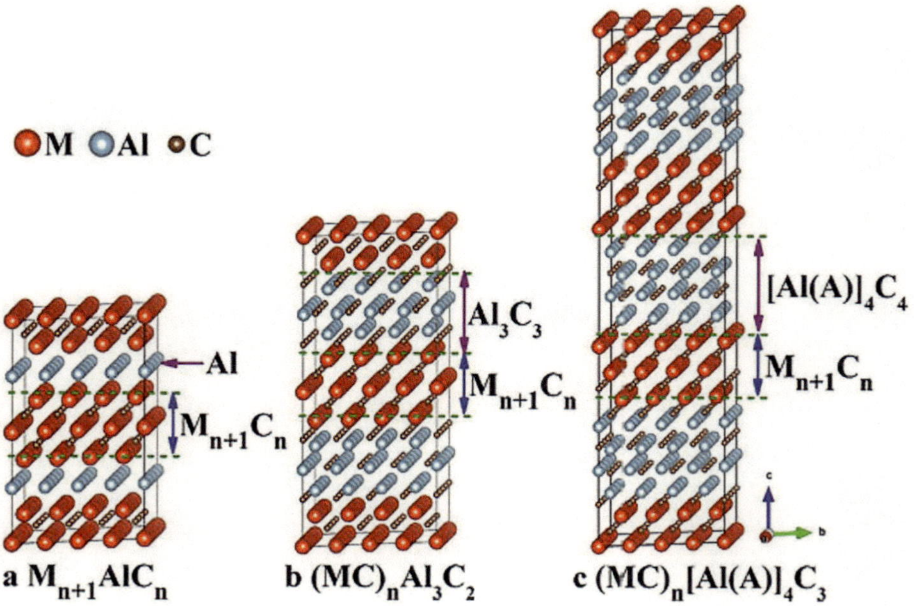

Figure 1.
Structure of the MAX phase (a), layered ternary (b), and quaternary (c) carbides of the type (MC)n[Al(A)] mCm⁻¹ (n = 1, 2, 3, m = 3, and 4) [8].

metals in the same compound has been introduced with increased stability for better thermal stabilization. The general formula can be written as $(M_xM'_{1-X})_{n+1}AX_n$, where the metal elements having similar electronic structure and atomic size are mixed to form the solution [12]. After the discovery of MXene, MAX phases have been investigated extensively. In 2014, a chemically ordered MAX phase was discovered with the structure of M_3AX_2 [13]. Chemically ordered MXene has highly organized arrangements of atoms. The well-defined position of M, A, and X atoms in the crystal lattice significantly impacts the material's properties. In the ordered MAX phase, the elements are in the ratio of 2:1 and form a phase that is non-coplanar, chemically ordered by altering the layers of the M element. Since the phase is non-coplanar, it is known as the 'out-of-plan' MAX phase, or 'o-MAX phase', and the MXene synthesized from it is known as 'o-MXene'. The first synthesized o-MAX phase was $(Cr_{2/3}Ti_{1/3})_3AlC_2$ (312) from Cr_2AlC and TiC at a molar ratio of 2:1 (Cr:Ti) [13]. The MXene synthesized from the ordered MAX phase has a distinct composition, where the outer layers were employed by M_1 atoms and the inner layers were occupied by M_2 atoms [7]. After 3 years of the discovery of o-MAX phase, Tao et al. have synthesized MAX phase $(Mo_{2/3}Sc_{1/3})_2AlC$ (211)) ordered in 'in-plane' and known as i-MAX. The MXene synthesized from it is called i-MXene. It is ordered in such a manner that M_1 atoms are occupied in the hexagonal lattice and M_2 atoms are occupied in the center of the hexagon.

MXene can also be synthesized by non-MAX phases, whose composition contains metals such as Sc, Zr and Hf and two A layers forming layered ternary or quaternary carbides. The structure can be described by layers of metal carbides $(M_{n+1}AlC_n)$ and aluminum carbides [8]. The non-MAX phases, like MoInC [14]. The figure (**Figure 2**) shows structures of MXenes from its initial discovery in 2011, followed by different phases, out-and-in-plan MXene and high entropy MXene.

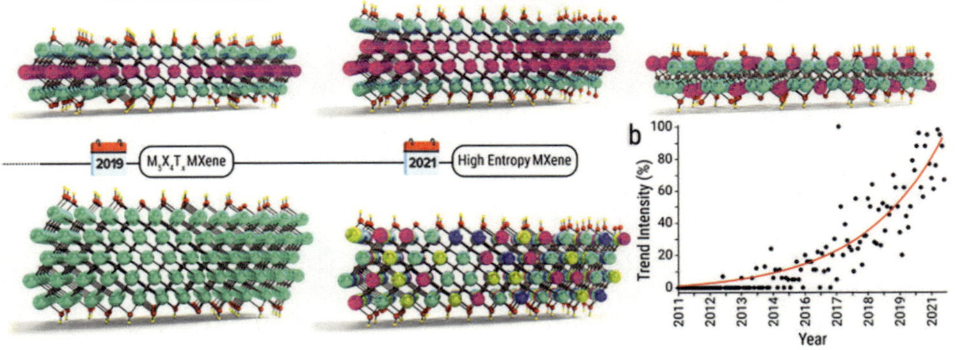

Figure 2.
(a) The image represents the different structures of MXenes from its initial discovery in 2011, followed by different phases, out-and-in-plan MXene and high entropy MXene. (b) Graph represents the trend data of the word MXene in Google, which shows its popularity and importance in the research field [29].

2.2 Synthesis methods

The primary synthesis approaches of MXene are selective etching and chemical vapor deposition (CVD), which are generally termed top-down and bottom-up approaches. Among them, the top-down approach: selective etching is widely used because of the availability of various combinations of bulk layered predecessors (MAX phases) of carbide and nitride. The bottom-up approach: CVD is known for its quality and precision.

Top-down; Selective etching: In selective etching, the layers of the precursor material are selectively etched using various etchants such as acid, alkali, fluoride + salt, molten salt, and NH_4HF_2 [7]. During the process, the layered MAX phase is soaked in the etchant for a long period, which weakens the bond between M–A and forms loosely bonded MXene. For example, in fluoride-based acid etching, the acid reacts with the A layer and forms AF_x with the liberation of hydrogen gas, then the $M_{n+1}X_n$ will again react with hydrofluoride and form $M_{n+1}X_nF_x$ [15]. In the above reaction, the functional group can be replaced with –O, –OH, –F, –Cl, –Br, etc. by using different etchants. Further, the material is delaminated by sonication to get a thin and monolayer of MXene. The functional groups mostly occupy the surface of the outermost layer of MXene, which drives the surface to attain negative charge and favors the development of stable colloidal dispersions [16]. This is known as surface termination; it enhances thermodynamic stability and other selective properties compared to pristine MXene. Also, during the process, various intercalants were used to speed up the process and to control the interlayer space by intercalating with different sizes of intercalants. Most used intercalants are organic molecules like dimethyl sulfoxide (DMSO) and tetrapropylammonium hydroxide and ions like Li^+, Na^+, K^+, Be^{2+}, Ca^{2+}, Mg^{2+}, etc.

The first MXene (Ti_3C_2) was synthesized in 2011 using hydrofluoric acid (HF) as an etchant. After the successful synthesis, numerous MXenes were produced using HF, making it the most used etchant. The concentration of fluoride ions and etching duration are inversely proportional. Also, high concentration and a longer time period are required for larger n values. Larger the n values, greater the energy of M–A bonds, which makes it harder to break [4].

Bottom-up; chemical vapor deposition (CVD): It is the most frequently used bottom-up strategy due to its high quality and accuracy. Multilayer epitaxial layers are normally deposited under controlled conditions to produce MXene [17]. Initially, a layer was deposited in a very high vacuum utilizing a DC magnetron sputtering process with the MAX phase. Nonetheless, translucent $Ti_3C_2T_x$ MXene films were created by selectively etching the A layer using ammonium hydrofluoride. MXenes' surfaces functionalized for improved electrochemical characteristics [7].

2.3 Properties of MXene

MXene's numerous qualities, including its high tensile strength, flexibility, hydrophilicity, high conductivity, and customizable surface termination, give it enormous potential. Because of these characteristics, MXenes are very adaptable and promising materials for a variety of uses, including optoelectronics, catalysis, electronics, and energy storage. It is possible to adjust MXene's characteristics through functionalization, doping, compositional changes, interlayer spacing changes, and processing [18, 19].

The most important property of MXene is metallic conductivity. The conductivity can be tuned by altering the process of etching and surface termination during the process. Conductivity of up to 24,000 S cm^{-1} was reported by Zeraati et al. for $Ti_3C_2T_x$ synthesized by the "Evaporated-Nitrogen" Minimally Intensive Layer Delamination (EN-MILD) technique, in which the concentration of Li^+ and acid is continually raised [20]. When the layers of MXene are functionalized by groups like F, OH, O, etc., the material turns into a semiconductor with a shift in band gap [21]. Since the functionalization takes place mostly in the outermost layer of the material, the outer layer plays an important role in electronic properties when compared to the inner layer [22]. The groups containing fewer valence electrons get functionalized easily. Intercalating MXene layers with ions like Li^+ also increases the electrical conductivity of the material. Intercalation of ions will increase the charge carrier density and surface area by altering the interlayer spacing. MXenes also have good mechanical properties due to strong M–C and M–N bonds, which multiplies the elastic constant by twice when compared to MAX phase [23]. When it comes to Young's modules, the increase in the number of layers of MXene will reduce Young's modulus by proportionality [24].

MXenes have good optical properties because of its absorbance in the UV-Visible-NIR range [25]. It can be used for applications such as photocatalysis, photovoltaic, and optoelectronics. It also has a good transmittance value, which can be optimized by altering thickness [26]. Optical properties can also be altered by functionalizing the surface of MXene. It affects the absorbance and transmission percentage [27]. MXenes also have magnetic properties, which are useful for sensors, detectors, storage devices, magnetic resonance imaging (MRI), etc.

Electrochemical properties of MXenes are greatly dependent on electronic properties. Mostly all the bare (or) pristine MXenes are metallic, but after it gets functionalized, some become semiconductors [28]. It could be due to the influence of functional groups in the energy bandgap. The large surface area to volume ratio increases electrochemical interaction and attributes better energy storage performance.

3. MXenes as electrode materials in supercapacitors

MXene as an active electrode material was first reported by Naguib et al. MXene shows high electrical conductivity in the range of 2×10^5 Sm^{-1} [30]. The multilayered structure and interlayer distance attribute good contact with the electrolyte. Apart from its properties, the interlayer distance of MXene can also be tuned systematically. At first, the capacitive results of MXene were reported as 300 Fcm^{-3}, which is greater than that of graphene [10]. Numerous MXene-based composites have been developed for energy storage devices, but they exhibit a lack of stability and reveal controversies between batteries and supercapacitors. MXenes have been shown to exhibit capacitance values as high as 900 F g^{-1}, which is significantly higher than traditional carbon-based materials used in supercapacitors. Additionally, MXenes have excellent cyclic stability, with little degradation in performance even after thousands of charge-discharge cycles. MXenes have also shown promise as anodes in lithium-ion batteries. MXenes exhibited high lithium-ion storage capacity, good rate capability, and cycling stability [10]. For example, $Ti_3C_2T_x$ MXene exhibited a reversible capacity of 311 mAh g^{-1} at a current density of 100 mA g^{-1}, which is significantly higher than traditional graphite anodes. In 2017, the maximum capacitance of 568 Fcm^{-3} and power density of 19,470 mW cm^{-3} were achieved by Leiquiang Qin et al. for solution processable $Mo_{1.33}C$ and PEDOT:PSS [31]. The $Ti_3C_2T_x$ MXene has reached high volumetric and gravimetric capacitance up to ≈1500 F cm^{-3} and ≈400 F g^{-1} [6]. At present, it is predominantly studied for ion batteries. The maximum specific capacity achieved is 1199 mAh g^{-1} at a current rate of 0.5 A g^{-1} for 450 cycles. These results show the difficulty in coexistence of both high power and charge density. But it can be achieved by developing pseudocapacitive material for electrochemical energy storage devices.

3.1 MXene over other 2D materials

If we compare MXenes with other two-dimensional materials, it exhibits superiority over other materials for several reasons. The principal factors comprise electrical conductivity, increased surface area, hydrophilicity, and tunable properties. MXenes exhibit far higher conductivity than other materials, particularly metals. This minimizes internal resistance and facilitates effective charge transfer. The layered architecture of MXene provides an enhanced surface area, facilitating several active sites for ion transport and adsorption/desorption. This also leads to an increase in energy density. Hydrophilicity significantly influences solubility and the interactions of ions between the layers. This facilitates an increased influx of ions into the layers of MXene. Pre-intercalation of ions during synthesis is a prevalent phenomenon that occurs at specific concentrations of MAX and the etchant employed. The interaction of intercalated ions or ions during solvent etching with hydroxides leads to the production of hydrated ions with a large radius. Due to the hydrophilic nature of MXene, hydrated ions can be effectively intercalated, facilitating further delamination. The carbides and nitrides facilitate functionalization with redox groups, allowing for the optimization of the material's electrochemical performance.

3.2 Evaluation of capacitance and stability

Electrochemical characterization of a material can be discussed with various terms such as capacitance, resistance, cyclic stability, retention, etc. Evolution of these terms can be done using techniques like Cyclic Voltammetry (CV), galvanometric charge discharge (GCD), and Electrochemical Impedance Spectroscopy (EIS). Similar to other supercapacitor materials, MXene has also been examined using analogous approaches. The prepared electrodes are often evaluated using a three-electrode method, which provides insights into the material's behavior and properties. The electrode material is further analyzed using two-electrode experiments to determine its anodic/cathodic characteristics, as well as its power and energy density. Ultimately, the material can be employed to construct a device (cell) and evaluate the final features in conjunction with the anode and cathode. All the aforementioned configurations are employed to analyze CV, GCD, and EIS. Based on the opinion of experts, it is very common to see improper or misleading interpretations of electrochemical characteristics among young researchers. The usual mistakes, such as distinguishing between diffusive and capacitive limited processes, interpreting the electrochemical performance matrix, and misinterpretations of electrochemical impedance data [32]. In this section, we are going to discuss the evaluation of electrochemical parameters and experts' advice regarding it.

The first step to distinguish the charge storage mechanism is to analyze the cyclic voltammograms and GCD curves. Materials with Electric Double Layered Capacitor (EDLC) or capacitive behavior will exhibit rectangular-like CVs and triangular GCD curves, which implies the linearity of voltage vs. time. If the CV curve deviates from a rectangular shape and has oxidation and reduction peaks, then it is said to be

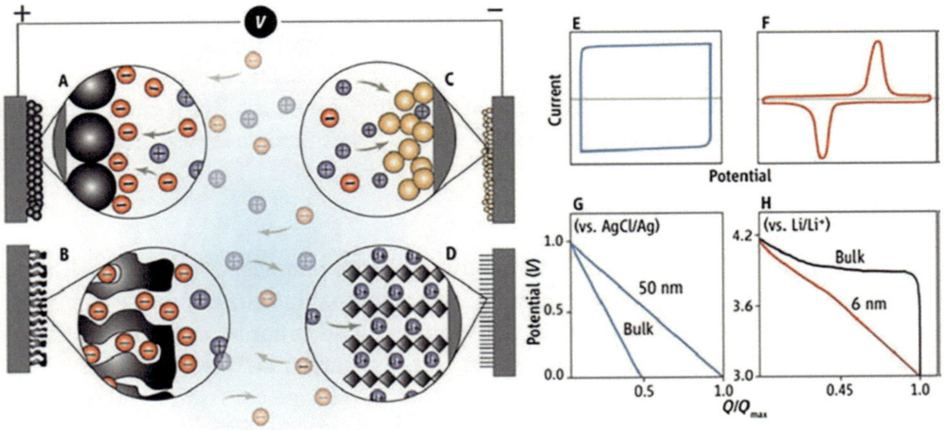

Figure 3
(A–D). *Mechanism of charge storage in capacitor. (A) and (B) EDLC behavior in carbon particles and porous carbon, where the negative ions are absorbed on the surface of positively charged electrodes. (C) Surface redox pseudocapacitor, like RuO_2, where the reaction takes place mostly over the surface. (D) Intercalation pseudocapacitance by Li^+. (E) and (F) CV curves of ideal capacitor and battery, i.e., capacitive and diffusive behavior. (G) and (H) discharge behavior of MnO_2 for nano and bulk material [50–53].*

diffusive or battery-like behavior [32]. In **Figure 3E** and **F** shows the ideal CV curves of EDLC and battery-type material.

The capacitance of a material is a positive constant of proportionality between charge and potential, which is denoted by *C* and described as the amount of charge stored in an electrode for the given potential range. The charging curve of first- and second-generation capacitors shows linear responses, but the new-generation capacitors, some EDLC, and pseudocapacitor show deviation from the linearity, which results in error or approximation in charge values. So, it is recommended to calculate the capacitance by integrating the charge over the potential range instead of approximation. For calculating the specific capacitance (C_{sc}) of the electrode, the value is divided by the mass (m) of the active material.

$$C_{sc} = \frac{1}{mVv}\int_{v_1}^{v_2} idV \tag{1}$$

where *V* is the potential range, ν is the scan rate, *i* is the discharge (or charge) current, and *dV* is change in potential. Capacitance is often reported as gravimetric capacitance (F g^{-1}) or volumetric capacitance (F ml^{-1}). In CV, fixing the potential window is very important; high or low potential causes the electrolyte to decompose [33]. At higher potential, the decomposition is oxidative and reductive at lower potential. For the galvanometric charge discharge curve, the equation for calculating capacitance is given by

$$C_{sc} = \frac{\Delta q}{\Delta V} = \frac{2i\int Vdt}{xV^2} \tag{2}$$

where *t* is discharging time, *V* is the potential range, *x* is the normalizing factor, and *i* is the discharge current. **Figure 4** shows the CV curves of pseudocapacitive behavior of MXene in various electrolytes and scan rates; *d* is the plot between scan rate vs. capacitance, called rate capability. Which shows an increase in scan rate

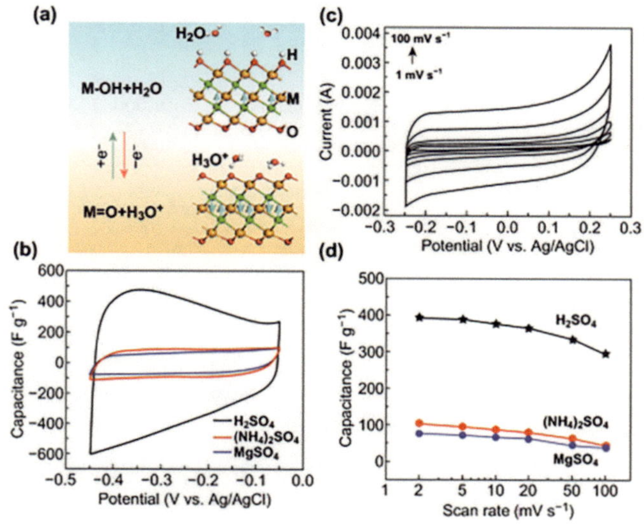

Figure 4.
The image depicts the pseudocapacitive performance of titanium carbide ($Ti_3C_2T_x$) MXene. (a) Interaction between electrolyte and MXene: change of surface group [44]. (b) CV plots of MXene at different electrolytes [45]. (c) CV plots of $Ti_3C_2T_x$ in different at various scan rates [46]. (d) Capacitance vs. scan rate for $Ti_3C_2T_x$ at different electrolytes [46]. (e) Illustration of pseudocapacitive interaction at solvated or desolated states (where the yellow is solvent molecule, green and orange are cation and anion) [47]. Reused under rights: original authors are cited as mentioned.

results in a decrease in capacitance because of less time for reaction. At a higher scan rate, the ions in the electrolyte do not have much time to penetrate through the pores, which results in less capacitance. Better rate capability means satisfactory specific capacitance even at higher scan rates.

For evaluating cyclic stability, GCD is performed for several cycles at a constant rate. Capacitive retention is the ratio between measured capacitance at each cycle and the initial cycle. As we mentioned above, higher and lower potential causes decomposition of electrolytes. The decomposed electrolyte particles occupy the pores of separation and activated carbon, which may increase the resistance during the cyclic process [34–36]. For evaluating capacitive retention, GCD is run for several cycles (~10,000,~5000) and the specific capacitance can be calculated for every 500 cycles and compared. Based on literature studies, non-Li-ion batteries can produce ~500 cycles, Na and Li-ion batteries can produce ~2000 cycles, supercapacitors deliver ~10,000 cycles, and hybrid systems can produce ~5000 cycles of lifespan. **Table 1** shows the comparison of specific capacitance and cyclic stability of various MXenes.

3.3 Charge storage mechanisms of supercapacitors

Supercapacitor charge storage mechanisms are broadly categorized into two types: Electric Double Layer Capacitor (EDLC) and pseudocapacitor. The EDLC stores energy through adsorption/desorption of charges on the electrode and electrolyte interface, which means the energy storage is done electrostatically without involving any chemical reaction. When voltage is applied, the charges (ions) will be attracted toward the opposite electrode and form an electric double layer between the electrode and electrolyte interface (**Figure 3A** and **B**). Whereas in a pseudocapacitor, the mechanism is dominated by faradaic redox reactions. This implies there will be

Material	Electrolyte	Specific capacitance	Cyclic stability	No. of cycles	Ref.
$Ti_3C_2T_x$	1 M KOH	340 F cm^{-3}	No degradation	10,000	[37]
d-$Ti_3C_2T_x$	1 M H_2SO_4	520 F cm^{-3}	No degradation	10,000	[38]
$V_4C_3T_x$	1 M H_2SO_4	~209 F g^{-1}	97.23%	10,000	[39]
$Nb_4C_3T_x$	1 M H_2SO_4	1075 F cm^{-3}	76%	5000	[40]
Ta_4C_3	0.1 M H_2SO_4	481 F g^{-1}	89%	2000	[41]
Mo_2CT_x	1 M H_2SO_4	700 F cm^{-3}	No degradation	10,000	[42]
Ni-$Mo_2TiC_2T_x$	3 M H_2SO_4	682 F g^{-1}	84%	5000	[43]

Table 1.
Comparison of specific capacitance and cyclic stability for supercapacitors derived of various MXenes.

reduction in anode and oxidation in cathode, which can be specified as charging and discharging. Charge storage occurs not only through redox reactions but also via charge separation at the interface. When potential is applied, the ions in the electrolyte will diffuse from the solution to the electrode/electrolyte interface. They will enter the bulk material (active material) through redox reactions, which are shown in **Figure 3C** and **D**. Comparatively, it allows the device to store more charge. During the discharge process, the stored ions will return to the electrolyte through a reverse redox reaction. The stored charges are released through an external circuit.

For example, if the redox active material is Ruthium oxide (RuO_x), then it will undergo a reduction reaction and form $uO_x\ \delta(OH)_z$. On the other hand, oxidation will take place in the positive electrode. The reaction can be written as (Eq. (3)).

$$RuO_x(OH)_z + \delta H + \delta e \leftrightarrow RuO_x\ \delta(OH)_z\ \delta \tag{3}$$

The redox reaction of the nonideal Faradaic process of titanium carbide $(Ti_3C_2T_x)$ in 1 M H_2SO_4 can be given as (Eq. (4)) [48].

$$Ti_3C_2O_2^q + 2H^+(aq) \rightarrow Ti_3C_2O_{2-2x}(OH)_{2x}^q + 2(1-x)H^+(aq) \tag{4}$$

where x = 0 to 1 (surface H coverage) and q = net charge on the electrode.

In intercalation pseudocapacitance, the dominant ions will penetrate the layers of the material during the redox reaction, which ultimately increases the participation of the ion during charge transfer. EDLC can produce high power density in the range of 5–20 kW kg^{-1} but lower energy density in the range of 1–10 Wh kg^{-1} [49]. Developing a pseudocapacitance material will increase the limit of energy and power density, which can be achieved by composite, doping, and hybrid devices.

4. MXene composites and hybrid materials

In recent times, the fabrication of composites has arisen as a compelling tactic for the development of enduring and versatile materials. Because of their dimensional morphology, layered structures, and excellent flexibility, MXenes have been widely recognized for their outstanding capability in the synthesis of multifunctional composites. As a result, there has been a significant increase in the number of studies centered on MXene-based composites. As of now, an array of special composite materials has been created through the combination of MXenes with various substances, such as polymers, metal oxides, and carbon nanotubes.

4.1 Various MXene composites

When combined with materials like polymers, carbon nanotubes, graphene, or nanoparticles, MXenes can generate composites with customized properties and improved functionalities. These composite materials find applications across fields including energy storage (supercapacitors and batteries), electronics (conductive materials), sensors (gas and biosensors), catalysis (efficient catalysts for diverse reactions), and structural components (lightweight and strong). The adaptability of MXene composites coupled with research and development efforts points to a future for these MXene composites in addressing technological challenges and driving advancements in emerging technologies. **Figure 5** shows the different kinds of MXene composites, such as carbon-based MXene composite, MXene conductive polymer composite, and MXene-metal composite.

By optimizing the interlayer spacing, MXene-based composites exhibit exceptional volumetric capacitance, pseudocapacitive behavior, excellent cycle stability, and high-rate capability. MXene composites have a unique property: the inclusion of properties such as robust electrical conductivity, clearly defined structure, and the ability to be modified has facilitated the achievement of high capacity and rapid charge-discharge capabilities in batteries. The fibers were stabilized and carbonized to produce MXene-carbon nanofibers. Conversely, the synthesis of MXene-conducting polymer composites involves the integration of

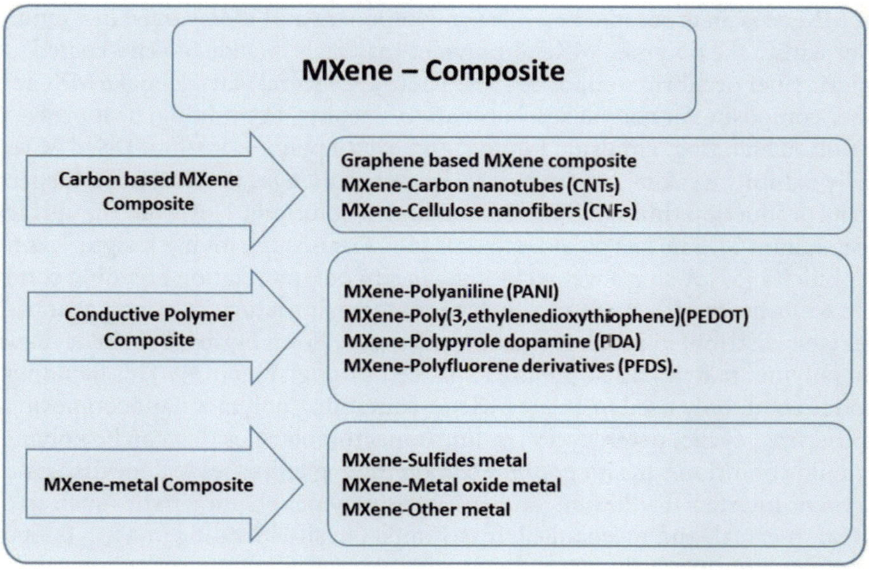

Figure 5.
Classification of various MXene composites.

defects in distinct MXene and conducting polymer materials, which are utilized in the fabrication of supercapacitor electrodes and exhibit remarkable mechanical properties. Compared to bare metal and MXene, MXene-metal displayed higher electrochemical supercapacitive performance; for example, in the case of co-ferrite and MXene, MXene/coexhibited superior electrochemical supercapacitive performance [54].

4.2 Synthesis of MXene composite

The addition of MXenes into material composites has been shown to notably improve the mechanical and thermal properties of the emerging materials. MXenes, known for their exceptional mechanical properties, hydrophilic surfaces, and metallic conductivity, offer significant advantages in these areas. In contrast to MXenes with many layers, MXenes with a single layer exhibit greater surface hydrophilicity and enhanced compatibility with the composite. Hence, it is generally accepted for carrying out the delamination of MXenes prior to their insertion in any materials such as polymer, carbon, or any oxide. It clearly shows the sandwich type of the combination of MXene with carbon nanotube (CNT) compounds results in significantly higher volumetric capacitance and rate performance compared to bare MXene. The incorporation of CNT into titanium carbide ($Ti_3C_2T_x$) increases the distance between the layers of $Ti_3C_2T_x$ flakes, allowing for the insertion of cations and enhancing the pathways for electrolyte ions. This leads to improved electrochemical performance of the $Ti_3C_2T_x$-CNT composite compared to the original MXene material. Moreover, the $Ti_3C_2T_x$-CNT is a type of MXene composite that can be used in lithium or sodium-ion batteries, and a strain sensor with exceptional properties, including remarkable sensitivity, high stretchability, and an adjustable sensing range, has been created [55].

Composites made of MXene and polymers frequently contain filled composites and complexes. In this instance, the term filled composite refers to a bulk material where MXene is incorporated as a filler or component and is dispersed in a random manner within the polymer. MXene/polymer materials include MXene-coated polymeric fiber or fabric composite [56]. There are several ways to make MXene/polymer composites or complexes: lamination stacking, dry mixing, heat pressing, solution blending, emulsion mixing, and in situ polymerization. Dried MXene typically exhibits weak hydrophilicity and poses challenges in terms of re-dispersion due to its production through fluoride-containing solutions. However, the surface of the resultant MXene has polar terminals of F, O, and OH, giving it significant hydrophilicity [57]. Using a wet technique, in situ polymerization blending combines MXene nanosheets with small-molecule monomers, initiators, or curing agents. This creates well-distributed macromolecules of MXene. From **Figure 6**, MXene dispersion in polymer matrix can be greatly enhanced through blending. This blending method is commonly used to create MXene-containing polymer nanocomposites with thermosetting, cyclic, heterocyclic, or linear macromolecules that can be polymerized under mild conditions. In situ polymerization mixing enhances MXene dispersion, strengthens interfacial adhesion, and enhances polymer characteristics such as electrical, thermal, and mechanical; for example, in situ blending of $Ti_3C_2T_x$/epoxy resin nanocomposites [58].

MXenes prevent the structural collapse of metal oxide and sulfide forms. There are four main ways to synthesize the composites, which include ex situ mixing, one-step in situ growth, multistep in situ conversion, and self-oxidation. Product structures and functions of MXene composites vary depending on the synthesis techniques

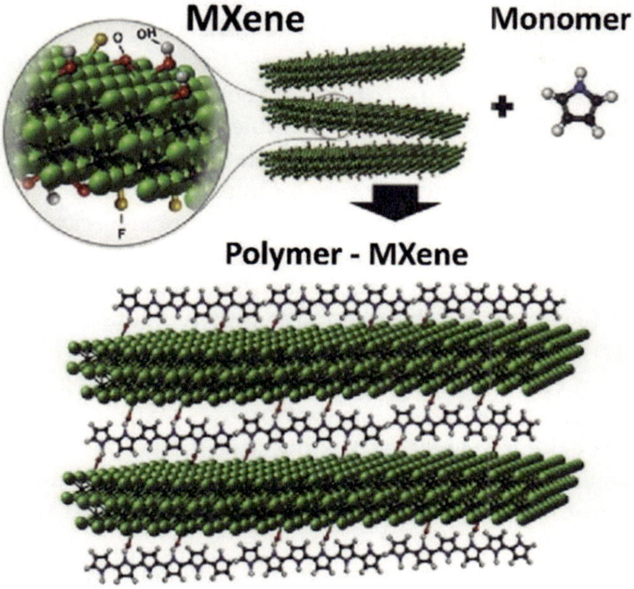

Figure 6.
Synthesis of MXene composites using the in situ method.

used. Stratified structures could be essential for generating more cavities for metal or metal compound deposits. Zhao et al. employed the ex situ method to manufacture structured $Ti_3C_2T_x/Co_3O_4$ composite, where the spray coating with two nozzles is used to supply MXene and metal oxide, allowing for control of layer thickness in the final material by altering spray time [59]. However, composites arranged differently require a basis (e.g., polymeric membrane) for support. Also, the high ionic conductivity of layered metal sulfides has spurred extensive investigation [60]. MXene/metal sulfide composites are typically synthesized using in situ hydrothermal methods.

4.3 MXene hybrid structures in supercapacitors

MXenes and its hybrid or composite structures have a variety of applications, but it is mostly targeted on energy storage applications because of its structure and electrochemical properties. By making a hybrid or composite material, the limitations of MXene, such as efficiency, electrical conductivity, reversible capacity, and limited active sites for reactions, are explored. Zheng et al. developed the MXene transition metal oxide hybrid anode, which consists of a conductive $Ti_3C_2T_x$ material that served as a bridge between the nanobelts of MoO_3. This anode achieved a maximum gravimetric energy density of 31.2 Wh kg^{-1} and a volumetric energy density of 39.2 Wh L^{-1} [61]. This can be attributed to the significant weight contribution of transition metal oxide. It also delivers good cyclic stability of approximately 94.2% after 10,000 cycles. Transition metal oxide MXene composites exhibit excellent gravimetric and volumetric capabilities, but their industrial scale performance is limited. Adding metal oxide to MXene also increases the pseudocapacitive performance of the material. The incorporation of a higher proportion of metal oxide into MXene can enhance pseudocapacitance to a specific extent. Zheng et al. developed a unique MXene aerogel structure that allows for homogeneously distributed MnO_2 nanosheets, resulting in a high pseudocapacitance near industrial standards. This unique structure provides 3D nucleation sites for covalently bonded MnO_2 nanosheets, resulting in a high energy density and power density. This hybrid structure offers superior performance and less mass loading, potentially extending to other MXene anode structures.

MXene composites can also be used for flexible and wearable electronic devices like sensors. Xie et al. developed a composite material, $CoFe_2O_4$ nanorod, incorporating metalorganic frameworks with MXenes for improved electrochemical performance and flexibility [62]. This material exhibits a volumetric capacitance of 2467.6 F cm^{-3} in LiCl electrolyte. It maintains stable electrochemical storage stability even when bent and retains 88.2% of its capacitance after 10,000 charge/discharge cycles. Muhammad et al. have reported that polar polymers containing charged nitrogen exhibit the highest affinity for MXenes, indicating the formation of effective hybrid pseudocapacitor structures [63]. This makes it a highly promising option for flexible energy storage devices in the future. The study reveals that charged nitrogen-containing polar polymers exhibit the highest binding strength with MXenes; this interaction enables the expansion of interlayer spacing and prevents restacking, indicating the possibility of incorporating organic materials into MXene layers to enhance hybrid pseudocapacitor structures [64].

Therefore, MXene hybrid composites can be experimentally synthesized with a focus on energy storage applications. It also explores electrode structure for improving supercapacitor applications; MXene-based heterostructures and hierarchical polymer structures are superior alternatives. The design of MXene hybrid anodes, pseudo capacitance in hierarchical structures [65], and the impact of electrolyte selection on supercapacitor energy storage have been investigated, highlighting the need of micro-supercapacitors for wearable energy storage. The review aims to help

MAX phase	MXene	Etchants	Applications	Ref.
Polymers				
Ti_3AlC_2	Ag (NW)/$Ti_3C_2T_x$ with PU	HCl/LiF	Strain sensor	[66]
Ti_3AlC_2	$Ti_3C_2T_x$@CS with PU	40% HF	Pressure sensor	[67]
Ti_3AlC_2	TiO_2-Ti_3C_2 with Nafion	HF	Bio sensor	[68]
Ti_3AlC_2	GOx/Au/$Ti_3C_2T_x$ with Nafion	—	Bio sensor	[69]
Carbon				
Ti_3AlC_2	Graphene/MXene hydrogel	HCl = LiF	Super capacitor	[70]
Ti_3AlC_2	VN/(PC) $Ti_3C_2T_x$	40% HF	Super capacitor	[71]
Ti_3AlC_2	MXene-graphene hydrogels	LiF + HCl	Super capacitor	[72]
Ti_3AlC_3	MXene@rGO	48%HF	Super capacitor	[73]
Oxides				
Ti_3AlC_2	$Ti_3C_2T_x$//PPy/MnO_2	LiF + HCl	Super capacitor	[74]
Ti_3AlC_2	N-MXene/TiO_2 heterostructure	LiF + HCl	Energy storage	[75]
Ti_3AlC_3	MnO_2@MXene/CNTF	LiF and HCl	Super capacitor	[76]
Ti_3AlC_2	MXene/CNT@MnO_2	LiF and HCl	Super capacitor	[77]

Table 2.
Various MXene composites and its applications (NW: nanowire; VN: vanadium nitride; PC: porous carbon; N-MXene: nitrogen-doped porous MXene).

identify suitable hybrid electrode composites and electrolyte selection for MXene-based supercapacitors. **Table 2** summarizes various MXene composites reported for different applications.

One of the key challenges in developing high-performance MXene-based energy storage devices is to design and fabricate electrodes with optimal nanostructures. A collection of nanostructured approaches has been developed to synthesize MXene electrodes with improved performance.

Hierarchical MXene structures are characterized by their multi-level composition, where smaller units combine to form larger entities, often exhibiting a synergy of properties. These structures can be tailored and precisely designed for specific applications, further expanding the potential of MXenes. The hierarchical MXene structures, such as nanosheets, nanowires, and other nanoarchitectures, represent a burgeoning area of research that holds immense potential for applications ranging from energy storage and conversion to electronics and beyond. As scientists delve deeper into the design and synthesis of these structures, the scope of MXene applications is expected to expand further, contributing to advancements in materials science and technology. The following section explores various approaches to enhance accessibility and charge transport of MXenes within their structures:

Surface functionalization: By changing the composition and structure of MXenes, it is possible to improve their electrochemical performance. For example, surface functionalization and elemental doping can increase the electrical conductivity and surface area of MXenes, which are both important for obtaining high power and energy density [78]. These tailored MXenes have more charge storage capacity and faster charge-discharge kinetics, making them an excellent choice for supercapacitors [79].

Interlayer spacing expansion: Increasing the separation distance between MXene layers can allow for more efficient ion diffusion and reduce ion blockages, enhancing charge transport and storage capabilities [80].

Doping and alloying: Doping or alloying MXenes with different elements can modify their electronic and structural properties, improving their charge transport properties and creating opportunities for tailored electrochemical performance [81].

Optimization of synthesis conditions: The choice of synthesis methods can significantly impact the ion accessibility and charge transport properties of MXenes. Researchers are continually optimizing synthesis conditions to produce MXenes with improved electrochemical performance [37].

Nanostructured MXene composites: To make MXene-based supercapacitors even better, we can combine them with other nanostructured materials, such as carbon nanotubes, graphene, or metal oxides. These new materials combine the best qualities of each material. They have a larger surface area, better electrical conductivity, and faster ion diffusion, all of which are important for high-performance supercapacitors [82].

Electrolyte optimization: The type of electrolyte used in a supercapacitor has a big impact on how long it lasts. Researchers have been working on optimizing the electrolyte composition to reduce side reactions and extend the lifespan of MXene-based supercapacitors [79].

Structural engineering: Researchers have developed new ways to structure MXene electrodes, such as core-shell architectures and protective coatings, to protect them from structural degradation during extended cycling [79].

5. Supercapacitor devices and prototypes

Supercapacitors that combine two different mechanisms of electrodes in a single system are called hybrid supercapacitors. Based on the arrangement, the hybrid supercapacitors can be classified into symmetric and asymmetric supercapacitors. The symmetric supercapacitors inherit electrodes of the same kind of material, whereas asymmetric supercapacitors consist of different kinds of electrodes (for example, capacitive and battery-type electrodes) [83, 84]. Combining electrodes with different mechanisms enhances the performance of the device; in combining a capacitive electrode with a battery-type electrode, the energy density and reversibility increase. The energy density range of Li-based systems is 120–200 Wh kg^{-1}. The power density range of supercapacitors is from 2 to 5 kW kg^{-1}. Pairing up these two types of electrodes will be an enhancing feature in hybrid supercapacitors [85].

MXene and MXene-based materials are used as electrodes in both symmetric and asymmetric supercapacitors. The symmetric supercapacitors are classified into 1D, 2D and 3D. MXene-based 1D supercapacitors are wire-type supercapacitors, which are known for flexibility and strength. It is mostly used in wearable electronic devices. Shao et al. synthesized electroactive nanofibers of MXene on textile polyester (PET) with high flexibility, strength, and power density. Also, various composites of nanofibers with MXene are synthesized, such as MXene/carbon nanofibers, MXene/carbon nanotube (CNT), etc. [86]. In 2D supercapacitors, the electrodes are mostly prepared by printing and coating. Also, the vacuum-filtered films of MXene/graphene and MXene/MnO_2 have directly been used as electrodes [87]. In MXene 3D supercapacitors composite stacks of layers, films such as MXene/carbon nanotubes and graphene are used. Materials having high porosity are used along with MXenes. The potential window of MXene as a negative electrode in an aqueous electrolyte is narrow, which limits energy density [86]. In an asymmetric supercapacitor, combining material with pseudocapacitive behavior (RuO_2) to MXene enhances the energy density with reliable power density. The asymmetric setup combining RuO_2 as a positive electrode and MXene as a negative electrode produces an energy density of 45 Wh cm^{-3} and a power density of 6 mW cm^{-2} [88]. Many asymmetric systems were accomplished by combining MXene cathode with rGO-polymer anode. Also, $Ti_3C_2T_x$//CoAl-LDH [89], AC//$Ti_3C_2T_x$/NiCoAl-LDH [90], and $Ti_3C_2T_x$/CuS/$Ti_3C_2T_x$ [91] are explored. Organic electrolytes containing Li^+, Na^+, and K^+ are used to attain broad potential windows. Lithiated material combining with MXene operates in a voltage range within 3 V. MXene is also explored with sodium-ion batteries. Most of the results of these systems are evident for fewer discharge cycles, so increasing cyclic stability to 5000–10,000 discharge cycles is a challenge. Apart from this, MXene-based transparent supercapacitors and microsupercapacitors have also been explored [86].

Developing a working supercapacitor involves various parameters, dimensions (thickness, size, and area) of the electrode and electrolyte, mechanisms of the active material, and design modules of the capacitor. Designing a supercapacitor mostly involves the alignment of the components, such as the electrode, electrolyte, separator, and sealant [95]. In aligning a supercapacitor, two electrodes (positive and negative) and an electrolyte are sandwiched by a separator in between. Commonly, the electrodes are in the thickness of ~100 μm for non-aqueous electrolytes and a bit thicker for aqueous electrolytes. The separator is in the thickness of ~24 μm and the current collector is ~50 μm [96]. In the aspect of alignment, the supercapacitors are commonly fabricated in three designs: coil cell, cylindrical cell, and pouch type cell, which is shown in **Figure 7**. In a coil cell, a single membrane of electrodes and

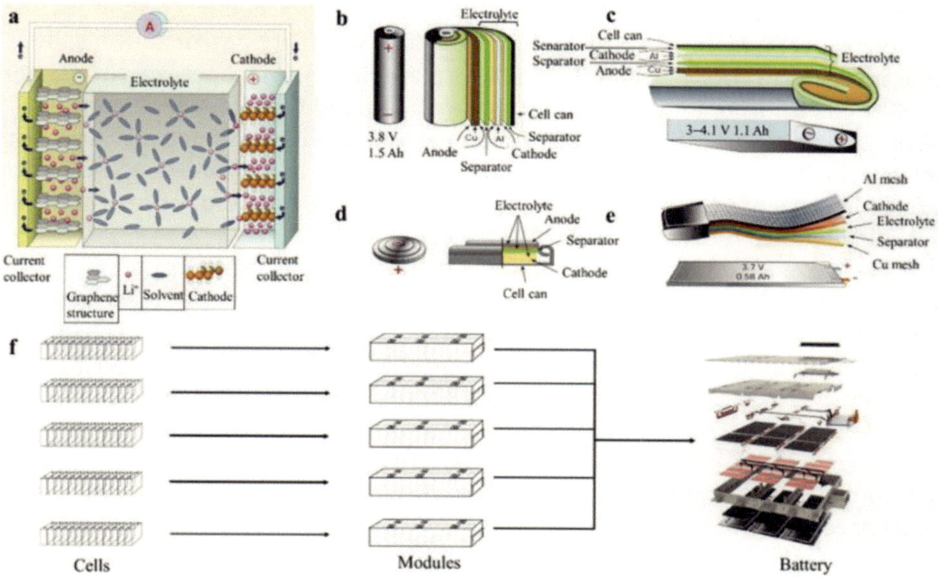

Figure 7.
Classification of supercapacitor cell design and arrangement (a); working mechanism of the supercapacitor cell (b); cylindrical type cell (c and e); pouch type cell (d); coil type cell (f) illustrate the arrangement of modules in the battery [92–94].

electrolyte is sandwiched and clamped by the cell caps. It delivers good performance on DC applications because of its thin dimension and less active material [97]. In a cylindrical cell, a long film of electrodes and separators is sandwiched and rolled into cylindrical casting, then filled with electrolyte. It is constructed with high mass loading, and the terminals are connected to both ends. In a pouch-type cell, several units of electrodes and separators are aligned, and the terminals are soldered in series connection. The final system is covered with a polymer bag. It has low equivalent series resistance and better performance.

The fabrication involves the preparation of electrodes and electrolytes. For the preparation of electrodes, a slurry or solution is prepared by mixing the active material with binder. It can be hot-pressed into thin films or coated on a substrate. The common methods used for coating in making commercial prototypes are spraying, casting, and printing, which have good surface area coverage and are easy to coat on non-conventional surfaces [97, 98].

6. MXenes for flexible and wearable devices

The convergence of electronics and textiles has given rise to a new era of wearable technology, revolutionizing industries such as healthcare, fitness, and fashion. In this context, one of the most critical challenges is ensuring a reliable and efficient power source for these wearable devices. Conventional energy sources are not suitable due to its bulky and rigid structure [86]. This challenge has been overcome by using supercapacitors, which are flexible and can power flexible devices. They can seamlessly integrate into clothing, sensors, and other wearable devices, ensuring a continuous and reliable power supply. Among the various materials and technologies used in these supercapacitors, MXene is known for its physical and chemical properties like high surface area, high electrical conductivity, high volumetric capacitance, enormous mechanical strength, and flexibility that are suitable for supercapacitor materials [99]. The incorporation of MXenes into wearable and flexible supercapacitors has opened new possibilities for energy storage in wearable technology. This section includes the importance, advantages, and properties of MXene-based supercapacitors in flexible and wearable devices. Also, it describes some of the applications of MXene supercapacitors in the field of wearable devices.

A diverse range of properties of MXenes made it suitable for energy storage devices for flexible and wearable devices [17, 100]. Some of them are listed below.

Compact form factor: MXene-based supercapacitors are thin and lightweight, making them suitable for integration into wearables without adding bulk or weight. This compact form factor aligns with the unobtrusive design requirements of many wearable devices.

Quick charging: The high electrical conductivity of MXenes ensures that supercapacitors can charge rapidly. This feature is particularly valuable in wearables, where users expect quick access to energy.

Long cycle life: MXenes exhibit excellent cyclic stability, enabling the extended operational lifespan of wearables. This is crucial for devices that are intended for prolonged use.

Safety: Unlike lithium-ion batteries, MXene-based supercapacitors do not pose the same risk of thermal runaway or combustion. They offer a safer energy storage solution for wearables.

Customizability: MXenes can be tailored to meet specific design requirements. Researchers can optimize their composition and structure to achieve desired properties, ensuring flexibility in designing supercapacitors for diverse wearable applications.

MXene-based supercapacitors have found application across diverse fields, including wearable electronics, medical devices, transportation, energy storage devices, energy harvesting systems, aerospace, flexible electronics, and more. Within this context, our focus lies on wearable and flexible devices, with some of the key applications outlined below.

Wearable electronics: MXene-based supercapacitors are integrated into textiles for wearable electronics, enabling applications such as health monitoring and energy harvesting [99]. They are called smart textiles. Early detection of disease is particularly important to treat it properly. The wearable health monitoring devices can give early information regarding any unhealthy conditions. Supercapacitors can act as power-supplying systems for the devices. Being flexible, it can be easily integrated into textiles without any discomfort to the user.

Wearable gadgets: MXene supercapacitors are integrated into wearable gadgets such as smartwatches and wireless earbuds to enhance energy storage capacity [17]. Today, the primordial importance in the manufacture of such devices is size and flexibility. The challenge was to reduce the size of the energy sources. The supercapacitors helped achieve this goal.

Flexible electronics: MXene supercapacitors are integrated into flexible electronic devices, such as rollable displays and e-paper [99].

MXene-based supercapacitors have demonstrated versatility and potential in a wide array of applications, contributing to advancements in energy storage technology and enabling innovative solutions in various industries. Some examples of real-world applications of MXene-based supercapacitors are described to get knowledge of how these are integrating into wearable and flexible devices. Thus, the research on MXenes for flexible supercapacitors is on the way to further studies and commercialization.

MXene-based wearable physiological biosensor: A 3D-printed wearable physiological biosensor was functionalized through an MXene-based supercapacitor [101]. Using the instrument, physiological changes can be monitored. This is the first system reported as a wearable system for continuous and real-time physiological biosignal monitoring. The system consists of the following parts: triboelectric nanogenerators, highly sensitive pressure generators, and multifunctional circuitry. [101]. Triboelectric nanogenerators are systems that convert mechanical energy to electrical energy through the processes of triboelectrification (generation of static charge through the contact and separation of different materials) and electrostatic induction. In the considered system, static charges are generated through friction because of human motion. This triboelectric nanogenerator contains an MXene supercapacitor, which can store energy obtained by triboelectric action. Thus, this flexible supercapacitor can replace bulky energy sources for wearable devices. The present system can monitor physiological conditions like heartbeat, blood pressure, etc. [101]. **Figure 8** can summarize the working of a wearable physiological biosensor, which yields an output power of 816.6 mW m^{-2} and a sensitivity of 6.03 kPa^{-1} with a detection limit of 9 Pa and a response time of 80 ms.

The main drawback of conventional nanomaterials for flexible devices was its poor mechanical stability. Thus, it's worth studying the mechanical stability and performance of MXenes. While MXenes demonstrate superior performance over

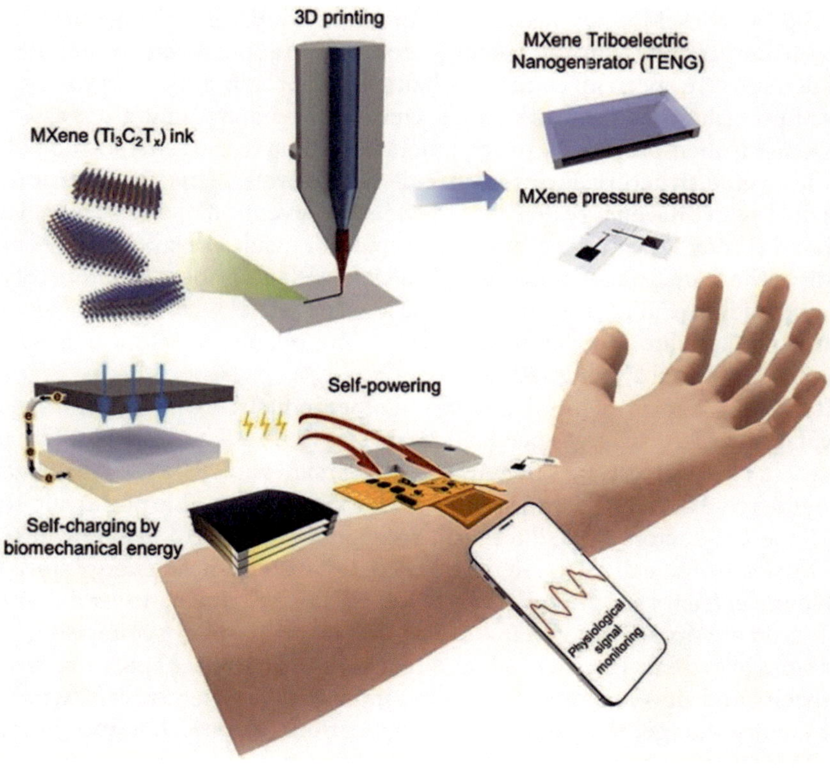

Figure 8.
Working of wearable sensor [101].

conventional materials in flexible devices, there is a demand for a novel and cost-effective fabrication process. Additionally, their susceptibility to oxidation upon contact with water poses a limitation to their practical applications. Use of MXene composites can address these problems. Ti_3C_2Tx is the only MXene extensively studied in this field.

There is a necessity to investigate and explore the properties of other MXenes for application in flexible devices [102]. The mechanical strengths of MXene depend on the surface terminations. It was observed that the mechanical strength of oxygen-terminated MXene is higher than that of the corresponding fluorine or hydroxyl termination. Though they are good materials for wearable electronics, certain factors raise challenges ahead [103] with the stacking problem of MXene, which limits the particle diffusion. The cost-effective and industrial production of MXenes has not yet been achieved. Also, we have no control over the surface terminations, and it is possible to synthesize MXenes of desired properties by suitably engineering the surface terminations.

7. Commercialization and industrial applications

The prospect of commercializing MXene materials for industrial applications is highly promising owing to their distinctive characteristics, including elevated electrical conductivity, exceptional thermal conductivity, robust mechanical strength, and remarkable ion intercalation capabilities. Additionally, they could include various mono- and multi-valent cations through intercalation, hence improving their electrochemical properties. MXenes possess the capacity to significantly transform diverse sectors, encompassing electronics, energy storage, water purification, and other related domains. To speed up commercializing MXenes, it is important to undertake a series of procedures and considerations. One of the primary limitations associated with MXenes is their propensity to agglomerate, leading to the re-deposition of ions inside the layered structure during charge-discharge cycles. This phenomenon can potentially restrict the energy density of MXenes; prevention of this occurrence can be achieved through the incorporation of a separator inside the layers of MXene, resulting in the formation of composites based on MXene. Supercapacitors (SCs) have found widespread utilization across various industrial sectors, and a significant amount of research literature exists that offers comprehensive technical insights into MXene-SC-integrated systems [104].

MXene supercapacitors (SCs) are used in more industries due to their high power density, fast charge/discharge speed, and long service life. They reduce transient electricity in power electronics like AC-DC conversion and DC-AC transformation and provide reliable electricity to critical loads in uninterruptible power supply (UPS). Superconductors can reduce ratio-frequency and electromagnetic interference as lossless voltage dividers. Renewable energy systems need energy storage to smooth out electricity supply and demand. To improve performance and reduce space, hybrid energy storage systems use complementary energy storage devices. Energy storage systems like fuel cells and rechargeable batteries benefit from SCs' high capacity and slow reaction rates. Other industries like electric cars, wind turbine energy storage, electronics, pulse applications, electrical motors, aerospace, and defense use supercapacitors and more. With continued advancements, supercapacitors hold the promise of powering grids and electric cars, underscoring their substantial industrial potential.

Furthermore, given that MXene supercapacitors (SCs) are consistently utilized to fulfill high-power requirements, it is imperative to develop accurate thermal models to facilitate effective thermal management design and ensure operational safety. There are different models utilized for simulating electrical behavior. These models are categorized into four distinct groups based on their structure, complexity, and accuracy: electrochemical models, equivalent circuit models, intelligent models, and fractional-order models. Extensive research has been devoted to investigating the mechanisms and developing robust models for simulating self-discharge. Diffusion and charge redistribution have been identified as the primary factors contributing to this phenomenon. In the context of thermal modeling, first principal models and comprehensive models for predicting the temperature of supercapacitors have been analyzed. The former demonstrates superior modeling precision, albeit at the expense of simplicity and computational speed. Additionally, taken into consideration are the techniques for calculating the state of charge (SOC) and tracking the state of health (SOH) of supercapacitors. The extended Kalman filter's self-correcting qualities and online implementation make it a popular choice in the literature for predicting state of charge (SOC). The survey indicates that most approaches begin by looking at the basic processes of aging in the context of state of health (SOH) monitoring. This emphasizes the impact of temperature, voltage, and cycling conditions on the aging process of supercapacitors. Subsequently, a multitude of SOH models have been developed in accordance with an analysis of aging mechanisms. Systems for power electronics, renewable energy sources, hybrid energy storage, and UPS are a few of the industrial uses for supercapacitors. Significant emphasis is placed on the supercapacitor's excellent power density and remarkable endurance in the context of these applications.

8. Environmental and sustainability aspects

In accordance with the environmental aspects, the MXenes are a sustainable and eco-friendly material, derived from naturally abundant and non-toxic sources like transition metal carbides or nitrides. MXenes are deemed to be low in toxicity, mitigating the risk of harmful byproducts during operation, which renders them well-suited for electronic devices that may come into contact with humans. They also offer high energy efficiency and rapid charge-discharge capabilities, reducing energy consumption and lowering environmental impacts. MXene-based supercapacitors have long cycle lives, reducing waste and the need for frequent manufacturing. Overall, MXenes offer a sustainable and eco-friendly alternative to traditional chemical processes. When we discuss the reversibility aspect, the MXene-based supercapacitor materials are easy to separate from other electronic device components, facilitating efficient recycling processes. Recovered materials can be reused in new supercapacitors or other electronic devices, maintaining their electrochemical performance even after recycling. This contributes to a sustainable supply chain for electronic components, aligning with the principles of a circular economy, where resources are efficiently used and reused, reducing the need for constant production of new materials. The eco-friendliness and recyclability of MXene-based supercapacitor materials are influenced by factors like the synthesis method, electrode material choice, and recycling processes. Researchers and industries are working to optimize these aspects to enhance the environmental benefits of MXene-based supercapacitors. MXene production requires careful consideration of environmental and ethical aspects. The abundance of transition metal carbides, nitrides, or carbonitrides in the Earth's crust is a significant factor in reducing resource scarcity. Many of these materials are non-toxic, ensuring environmental sustainability and human safety. Recycling and reusing source materials can reduce the need for additional mining and extraction. Waste minimization is another important aspect, involving reducing hazardous chemicals and optimizing processing methods. Efficient resource use is essential, with methods that maximize yield and minimize waste. A life cycle assessment evaluates the environmental impact from extraction to disposal. Sustainable practices can reduce the carbon footprint associated with source material extraction and processing, including using renewable energy sources and minimizing transportation-related emissions. Future research and development efforts should focus on improving sustainability, including the development of eco-friendly synthesis techniques. Collaborations between research institutions, industry, and regulatory bodies can establish standards and certifications for sustainable MXene extraction and processing. MXenes are eco-friendly, recyclable, and non-toxic, contributing to overall energy efficiency. They can be customized to specific applications, enhancing energy utilization. While not replacing traditional technologies, MXenes complement existing systems and offer advantages in power delivery and sustainability.

9. Conclusion and future outlook

We have discussed the basics of MXene material and investigated its electrochemical characteristics, including charge storage mechanisms. The pseudocapacitive behavior and tunability of layers make it an indisputable material for supercapacitors. By intercalation of ions, the performance and cyclability can be improved by increasing the redox reaction. Also, MXene-based composites and hybrid materials are studied for future devices. Devices like wearable supercapacitors, prototypes, and its parameters were discussed. Further, industrialization and commercialization of supercapacitors with their needs are examined.

Based on the investigation, we can see the prospects of MXene. Even though it has certain limitations, its potential in the field of energy storage devices is quite clear. The pseudocapacitive behavior of MXene and MXene-based composites, along with layered structure, will be a key feature for enhancement in performance. Ion intercalation in between the layers of MXene directly influences the electrochemical behavior and performance. In some cases, addition of excess ions results in decreases in interatomic layer due to interaction between MXene layers and hydrated ions. Hence, by controlling ion intercalation, it is possible to tailor the structural characteristics of MXene. Also, utilizing selective ionic electrolytes will improve the charge kinetics and improve electrochemical performance.

Synthesizing composites and fabricating hybrid systems are the prominent ways to improve the performance of the material. Fabricating hybrid devices by combining MXene electrodes with materials that have alternative behavior will balance and enhance the performance, i.e., balanced energy and power density with better cyclic stability. The dimensions and parameters are very important for developing a prototype and commercializing it. In most of the research works, the prototype systems are designed with custom parameters, which makes it hard to compare with other and market devices. Additionally, adjusting parameters can yield better performance in laboratory tests but often fails to provide reliable results in real-world applications. So, it is highly recommended to follow the commercial parameters as discussed. The flexibility of the MXene material makes it much more suitable for supercapacitors in wearable sensors.

In conclusion, we explore MXene's potential as a supercapacitor material, emphasizing its pseudocapacitive behavior and tunable layered structure, which enhance its energy storage capabilities. We found that ion intercalation significantly boosts performance and cycling stability, although precise control is necessary to avoid reduced interlayer spacing. The fine-tuning of MXene's electrochemical performance through selective ionic electrolytes opens new avenues for improving charge kinetics and storage efficiency. MXene-based composites and hybrid materials offer a balanced approach to achieving high energy density, power density, and cyclic stability, especially relevant for flexible, wearable sensor applications. Ultimately, we also identify challenges in standardizing prototype designs, stressing the importance of adhering to commercial parameters for real-world scalability. It's exciting to see MXene's potential in the world of wearable technology, and further research is required to delve deeper into the subject and expand our understanding of the field.

Acknowledgements

Gopi R R, Amritha Kadathamana, and Jithin Varghese to Conahcyt-Mexico for providing a research scholarship.

Conflict of interest

The authors declare no conflict of interest.

Appendices and nomenclature

Charge density	the amount of electric charge stored in a given volume or mass of material
Charge-discharge cycles	the process of charging a battery or capacitor and then discharging it to release energy
Csc	specific capacitance, measured in Farads per gram (F g^{-1}), indicating the charge storage capacity of a supercapacitor normalized to its mass
Current density	the electric current per unit area of cross-section, often expressed in mA g^{-1} or A g^{-1} for electrode materials
CV	cyclic voltammetry, an electrochemical technique used to evaluate the electrochemical properties of a material by cycling the voltage and measuring current
CVD	chemical vapor deposition, a bottom-up synthesis approach known for quality and precision
Cyclic stability	the ability of a material to maintain performance over repeated charge and discharge cycles

DMSO	dimethyl sulfoxide, an organic molecule often used as an intercalant
EIS	electrochemical impedance
Electrical conductivity	the ability of a material to conduct electric current
EN-MILD	evaporated-nitrogen minimally intensive layer delamination, a technique for synthesizing MXenes with high conductivity
EVs	electric vehicles, vehicles powered by electric energy
$F\ g^{-1}$	Farads per gram, a unit of specific capacitance
$F\ ml^{-1}$	Farads per milliliter, a unit of volumetric capacitance
$F\ g^{-1}$	a unit of specific capacitance used to express the capacitance of a supercapacitor or capacitor material relative to its mass, indicating how much electrical charge it can store per gram of material
$F\ cm^{-3}$	denotes the capacitance density of a supercapacitor in Farads per cubic meter, indicating its energy storage capacity per unit volume
GCD	galvanometric charge discharge, a method to measure the capacity of a battery or supercapacitor by applying a constant current to charge and discharge the device
Graphene	a single layer of carbon atoms arranged in a two-dimensional honeycomb lattice, known for its excellent electrical properties
Hafnium (Hf)	a transition metal with atomic number 72, recognized for its high melting point and excellent neutron-capture properties, commonly used in nuclear reactors and as a material for high-temperature applications
HF	hydrofluoric acid, the etchant used to synthesize the first MXene (Ti_3C_2)
Intercalants	substances used to control the interlayer spacing in MXenes, including ions like Li^+, Na^+, K^+, Be^{2+}, Ca^{2+}, and Mg^{2+}
$kW\ kg^{-1}$	a unit of power density that measures the amount of power output per unit mass of a material or device, commonly used to assess the performance of energy storage systems such as batteries and supercapacitors
M–A bond	metallic bond between transition metals and group 13/14 elements in MAX phases
Magnetic resonance imaging (MRI)	a medical imaging technique used to visualize internal structures of the body, highlighting the potential use of MXenes in imaging technologies

$mAh\ g^{-1}$	milliampere-hours per gram, a unit of specific capacity for batteries
MAX Phase	a class of layered materials represented by the formula Mn + 1AXn, where A is typically a group 13 or 14 element
M–C and M–N bonds	bonds between transition metals (M) and carbon (C) or nitrogen (N) in MXenes, contributing to their mechanical properties
$mW\ cm^{-3}$	milliwatts per cubic centimeter, a unit of power density
M–X bond	ionic or covalent bond between transition metals and carbon/nitrogen in MXenes
MXene	a class of two-dimensional materials made from metal carbides or nitrides
P63/mmc symmetry	the hexagonal crystal symmetry typical of many MXenes and MAX phases
Power density	the amount of power (energy per unit time) that can be delivered by a device per unit volume or weight
Pseudocapacitors	a type of capacitor that stores energy through fast surface redox reactions
Pseudocapacitive material	materials that exhibit both electrochemical double-layer capacitance and faradaic charge transfer, contributing to high power and energy densities in supercapacitors
Rate capability	ability to quickly deliver stored energy at high discharge rates while maintaining efficiency and performance, making them suitable for applications requiring rapid bursts of power
Redox reactions	redox reactions refer to chemical processes involving the simultaneous oxidation and reduction of species, characterized by the transfer of electrons between reactants
$S\ cm^{-1}$	Siemens per centimeter, a unit of electrical conductivity
$S\ m^{-1}$	Siemens per meter, a unit of electrical conductivity
Scandium (Sc)	a transition metal with atomic number 21, known for its lightweight and strength, often used in alloys and as a doping agent in aluminum to enhance its properties
Supercapacitors	electrochemical devices that store energy through the electrostatic separation of charge
Surface area to volume ratio	a measure that influences the reactivity and performance of materials, especially in electrochemical applications
Surface termination (T)	the presence of functional groups on the surface of MXenes, influencing their electrochemical properties

UV-Vis-NIR	ultraviolet-visible-near infrared, a range of light that MXenes can absorb, relevant for their optical properties
Volumetric capacitance	the capacitance of a material per unit volume
$Wh{\cdot}kg^{-1}$	a unit of energy density that represents the amount of energy stored per unit mass of a material or device, often used to evaluate the energy storage capacity of batteries and supercapacitors
$Wh\ L^{-1}$	a unit of energy density that measures the amount of energy stored per unit volume of a material or device, commonly used to assess the energy storage capacity of batteries and supercapacitors
$Wh{\cdot}kg^{-1}$	a unit of energy density that expresses the amount of energy stored per unit mass of a material or device, often used to evaluate the performance of batteries and supercapacitors
Young's modulus	a measure of the stiffness of a solid material, defined as the ratio of stress to strain

Author details

Gopi R.R.[2], Amritha Kadathamana[2], Jithin Varghese[3], Shadai Lugo Loredo[1], Andrea Cerdán-Pasarán[1], Soorya Pushpan[4], J.A. Hernández-Magallanes[1] and Sanal K.C.[1]*

1 Facultad de Ciencias Químicas, Universidad Autónoma de Nuevo León, San Nicolas de los Garza, Nuevo León, Mexico

2 Facultad de Ingeniería Mecánica y Eléctrica, Universidad Autónoma de Nuevo León, San Nicolas de los Garza, Nuevo León, Mexico

3 Facultad de Ciencias Físico Matemáticas, Universidad Autónoma de Nuevo León, San Nicolas de los Garza, Nuevo León, Mexico

4 Tecnológico de Monterrey, Monterrey, Nuevo León, Mexico

*Address all correspondence to: sanal.kozhiparambilch@uanl.edu.mx

References

[1] Zhang H, Zhang J, Gao X, Wen L, Li W, Zhao D. Advances in materials and structures of supercapacitors. Ionics (Kiel). 2022;**28**:515-531. DOI: 10.1007/S11581-021-04359-5/METRICS

[2] Abdel Maksoud MIA, Fahim RA, Shalan AE, Abd Elkodous M, Olojede SO, Osman AI, et al. Advanced materials and technologies for supercapacitors used in energy conversion and storage: A review. Environmental Chemistry Letters. 2020;**19**:375-439. DOI: 10.1007/S10311-020-01075-W

[3] Ling Z, Ren CE, Zhao MQ, Yang J, Giammarco JM, Qiu J, et al. Flexible and conductive MXene films and nanocomposites with high capacitance. Proceedings of the National Academy of Sciences of the United States of America. 2014;**111**:16676-16681. DOI: 10.1073/PNAS.1414215111/SUPPL_FILE/PNAS.1414215111.SM02.WMV

[4] Kumar YA, Raorane CJ, Hegazy HH, Ramachandran T, Kim SC, Moniruzzaman M. 2D MXene-based supercapacitors: A promising path towards high-performance energy storage. Journal of Energy Storage. 2023;**72**:108433. DOI: 10.1016/J.EST.2023.108433

[5] Pang J, Mendes RG, Bachmatiuk A, Zhao L, Ta HQ, Gemming T, et al. Applications of 2D MXenes in energy conversion and storage systems. Chemical Society Reviews. 2019;**48**:72-133. DOI: 10.1039/C8CS00324F

[6] Mohammadi AV, Rosen J, Gogotsi Y. The world of two-dimensional carbides and nitrides (MXenes). Science. 2021;**372**. DOI: 10.1126/SCIENCE.ABF1581/ASSET/81125F49-2018-4AA2-BB0D-1C892C90A287/ASSETS/GRAPHIC/372_ABF1581_F6.JPEG

[7] Pogorielov M, Smyrnova K, Kyrylenko S, Gogotsi O, Zahorodna V, Pogrebnjak A. MXenes—A new class of two-dimensional materials: Structure, properties and potential applications. Nanomaterials. 2021;**11**:3412. DOI: 10.3390/NANO11123412

[8] Zha XH, Zhou J, Eklund P, Bai X, Du S, Huang Q. Non-MAX phase precursors for MXenes. In: 2D Metal Carbides and Nitrides (MXenes): Structure, Properties and Applications. Cham, Switzerland AG: Springer; 2019. pp. 53-68. DOI: 10.1007/978-3-030-19026-2_4/COVER

[9] Jimmy J, Kandasubramanian B. MXene functionalized polymer composites: Synthesis and applications. European Polymer Journal. 2020;**122**:109367. DOI: 10.1016/J.EURPOLYMJ.2019.109367

[10] Naguib M, Come J, Dyatkin B, Presser V, Taberna PL, Simon P, et al. MXene: A promising transition metal carbide anode for lithium-ion batteries. Electrochemistry Communications. 2012;**16**:61-64. DOI: 10.1016/J.ELECOM.2012.01.002

[11] Barsoum M, El-Raghy T. The MAX phases: Unique new carbide and nitride materials. American Scientist. 2001;**89**:334. DOI: 10.1511/2001.28.736

[12] Persson I, el Ghazaly A, Tao Q, Halim J, Kota S, Darakchieva V, et al. Tailoring structure, composition, and energy storage properties of MXenes from selective etching of in-plane, chemically ordered MAX phases. Small. 2018;**14**:1703676. DOI: 10.1002/SMLL.201703676

[13] Liu Z, Wu E, Wang J, Qian Y, Xiang H, Li X, et al. Crystal

structure and formation mechanism of $(Cr_2/3Ti_{1/3})_3AlC_2$ MAX phase. Acta Materialia. 2014;**73**:186-193. DOI: 10.1016/J.ACTAMAT.2014.04.006

[14] Thomas T, Pushpan S, Aguilar Martínez JA, Torres Castro A, Pineda Aguilar N, Álvarez-Méndez A, et al. UV-assisted safe etching route for the synthesis of Mo_2CT_x MXene from Mo-In-C non-MAX phase. Ceramics International. 2021;**47**:35384-35387. DOI: 10.1016/J.CERAMINT.2021.08.342

[15] Wu Y, Nie P, Wang J, Dou H, Zhang X. Few-layer MXenes delaminated via high-energy mechanical milling for enhanced sodium-ion batteries performance. ACS Applied Materials & Interfaces. 2017;**9**:39610-39617. DOI: 10.1021/ACSAMI.7B12155/SUPPL_FILE/AM7B12155_SI_001.PDF

[16] Naguib M, Saito T, Lai S, Rager MS, Aytug T, Parans Paranthaman M, et al. $Ti_3C_2T_x$ (MXene)–polyacrylamide nanocomposite films. RSC Advances. 2016;**6**:72069-72073. DOI: 10.1039/C6RA10384G

[17] Anasori B, Lukatskaya MR, Gogotsi Y. 2D metal carbides and nitrides (MXenes) for energy storage. Nature Reviews Materials. 2017;**2**:1-17. DOI: 10.1038/natrevmats.2016.98

[18] Sun W, Wang HW, Vlcek L, Peng J, Brady AB, Osti NC, et al. Multiscale and multimodal characterization of 2D titanium carbonitride MXene. Advanced Materials Interfaces. 2020;**7**:1902207. DOI: 10.1002/ADMI.201902207

[19] Zhu X, Liu P, Xue T, Ge Y, Ai S, Sheng Y, et al. A novel graphene-like titanium carbide MXene/Au–Ag nanoshuttles bifunctional nanosensor for electrochemical and SERS intelligent analysis of ultra-trace carbendazim coupled with machine learning. Ceramics International. 2021;**47**:173-184. DOI: 10.1016/J.CERAMINT.2020.08.121

[20] Shayesteh Zeraati A, Mirkhani SA, Sun P, Naguib M, Braun PV, Sundararaj U. Improved synthesis of $Ti_3C_2T_x$ MXenes resulting in exceptional electrical conductivity, high synthesis yield, and enhanced capacitance. Nanoscale. 2021;**13**:3572-3580. DOI: 10.1039/D0NR06671K

[21] Naguib M, Kurtoglu M, Presser V, Lu J, Niu J, Heon M, et al. Two-dimensional nanocrystals produced by exfoliation of Ti_3AlC_2. Advanced Materials. 2011;**23**:4248-4253. DOI: 10.1002/ADMA.201102306

[22] Anasori B, Xie Y, Beidaghi M, Lu J, Hosler BC, Hultman L, et al. Two-dimensional, ordered, double transition metals carbides (MXenes). ACS Nano. 2015;**9**:9507-9516. DOI: 10.1021/ACSNANO.5B03591/ASSET/IMAGES/NN-2015-03591M_M002.GIF

[23] Kulkarni AA, Gaikwad NK, Salunkhe AP, Dahotre RM, Bhat TS, Patil PS. 2D MXene integrated strategies: A bright future for supercapacitors. Journal of Energy Storage. 2023;**71**:107975. DOI: 10.1016/J.EST.2023.107975

[24] Wyatt BC, Rosenkranz A, Anasori B. 2D MXenes: Tunable mechanical and tribological properties. Advanced Materials. 2021;**33**:2007973. DOI: 10.1002/ADMA.202007973

[25] Wang J, Cai Z, Lin D, Chen K, Zhao L, Xie F, et al. Plasma oxidized $Ti_3C_2T_x$ MXene as electron transport layer for efficient perovskite solar cells. ACS Applied Materials & Interfaces. 2021;**13**:32495-32502. DOI: 10.1021/ACSAMI.1C07146/SUPPL_FILE/AM1C07146_SI_001.PDF

[26] Hantanasirisakul K, Zhao MQ, Urbankowski P, Halim J, Anasori B, Kota S, et al. Fabrication of $Ti_3C_2T_x$ MXene transparent thin films with tunable optoelectronic properties. Advanced Electronic Materials. 2016;**2**:1600050. DOI: 10.1002/AELM.201600050

[27] Berdiyorov GR. Optical properties of functionalized $Ti_3C_2T_2$ (T = F, O, OH) MXene: First-principles calculations. AIP Advances. 2016;**6**:55105. DOI: 10.1063/1.4948799/995547

[28] Panda S, Deshmukh K, Khadheer Pasha SK, Theerthagiri J, Manickam S, Choi MY. MXene based emerging materials for supercapacitor applications: Recent advances, challenges, and future perspectives. Coordination Chemistry Reviews. 2022;**462**:214518. DOI: 10.1016/J.CCR.2022.214518

[29] Dadashi Firouzjaei M, Karimiziarani M, Moradkhani H, Elliott M, Anasori B. MXenes: The two-dimensional influencers. Materials Today Advances. 2022;**13**:100202. DOI: 10.1016/J.MTADV.2021.100202

[30] An H, Habib T, Shah S, Gao H, Radovic M, Green MJ, et al. Surface-agnostic highly stretchable and bendable conductive MXene multilayers. Science Advances. 2018;**4**:1-8. DOI: 10.1126/SCIADV.AAQ0118/SUPPL_FILE/AAQ0118_SM.PDF

[31] Qin L, Tao Q, El Ghazaly A, Fernandez-Rodriguez J, Persson POÅ, Rosen J, et al. High-performance ultrathin flexible solid-state supercapacitors based on solution processable Mo1.33C MXene and PEDOT:PSS. Advanced Functional Materials. 2018;**28**:1703808. DOI: 10.1002/ADFM.201703808

[32] Mathis TS, Kurra N, Wang X, Pinto D, Simon P, Gogotsi Y. Energy storage data reporting in perspective—Guidelines for interpreting the performance of electrochemical energy storage systems. Advanced Energy Materials. 2019;**9**:1902007. DOI: 10.1002/AENM.201902007

[33] Uno M, Tanaka K. Accelerated charge-discharge cycling test and cycle life prediction model for supercapacitors in alternative battery applications. IEEE Transactions on Industrial Electronics. 2012;**59**:4704-4712. DOI: 10.1109/TIE.2011.2182018

[34] German R, Venet P, Sari A, Briat O, Vinassa JM. Comparison of EDLC impedance models used for ageing monitoring. In: Proceedings of the 2012 1st International Conference on Renewable Energies and Vehicular Technology, REVET 2012. Nabeul, Tunisia: IEEE; 2012. pp. 224-229. DOI: 10.1109/REVET.2012.6195275

[35] El Brouji EH, Briat O, Vinassa JM, Bertrand N, Woirgard E. Impact of calendar life and cycling ageing on supercapacitor performance. IEEE Transactions on Vehicular Technology. 2009;**58**:3917-3929. DOI: 10.1109/TVT.2009.2028431

[36] Hahn M, Kötz R, Gallay R, Siggel A. Pressure evolution in propylene carbonate based electrochemical double layer capacitors. Electrochimica Acta. 2006;**52**:1709-1712. DOI: 10.1016/J.ELECTACTA.2006.01.080

[37] Lukatskaya MR, Mashtalir O, Ren CE, Dall'Agnese Y, Rozier P, Taberna PL, et al. Cation intercalation and high volumetric capacitance of two-dimensional titanium carbide. Science. 2013;**341**:1502-1505. DOI: 10.1126/SCIENCE.1241488/SUPPL_FILE/LUKATSKAYA.SM.PDF

[38] Dall'Agnese Y, Lukatskaya MR, Cook KM, Taberna PL, Gogotsi Y,

Simon P. High capacitance of surface-modified 2D titanium carbide in acidic electrolyte. Electrochemistry Communications. 2014;**48**:118-122. DOI: 10.1016/J.ELECOM.2014.09.002

[39] Wang X, Lin S, Tong H, Huang Y, Tong P, Zhao B, et al. Two-dimensional V_4C_3 MXene as high performance electrode materials for supercapacitors. Electrochimica Acta. 2019;**307**:414-421. DOI: 10.1016/J.ELECTACTA.2019.03.205

[40] Zhao S, Chen C, Zhao X, Chu X, Du F, Chen G, et al. Flexible $Nb_4C_3T_x$ film with large interlayer spacing for high-performance supercapacitors. Advanced Functional Materials. 2020;**30**:2000815. DOI: 10.1002/ADFM.202000815

[41] Syamsai R, Grace AN. Ta_4C_3 MXene as supercapacitor electrodes. Journal of Alloys and Compounds. 2019;**792**:1230-1238. DOI: 10.1016/J.JALLCOM.2019.04.096

[42] Halim J, Kota S, Lukatskaya MR, Naguib M, Zhao MQ, Moon EJ, et al. Synthesis and characterization of 2D molybdenum carbide (MXene). Advanced Functional Materials. 2016;**26**:3118-3127. DOI: 10.1002/ADFM.201505328

[43] Hakim MW, Fatima S, Tahir R, Iqbal MZ, Li H, Rizwan S. Ni-intercalated $Mo_2TiC_2T_x$ free-standing MXene for excellent gravimetric capacitance prepared via electrostatic self-assembly. Journal of Energy Storage. 2023;**61**:106662. DOI: 10.1016/J.EST.2023.106662

[44] Tang Y, Zhu J, Yang C, Wang F. Enhanced capacitive performance based on diverse layered structure of two-dimensional Ti_3C_2 MXene with long etching time. Journal of the Electrochemical Society. 2016;**163**:A1975-A1982. DOI: 10.1149/2.0921609JES/XML

[45] Hu M, Li Z, Hu T, Zhu S, Zhang C, Wang X. High-capacitance mechanism for $Ti_3C_2T_x$ MXene by in situ electrochemical Raman Spectroscopy investigation. ACS Nano. 2016;**10**:11344-11350. DOI: 10.1021/ACSNANO.6B06597/SUPPL_FILE/NN6B06597_SI_001.PDF

[46] Gao Y, Wang L, Li Z, Zhang Y, Xing B, Zhang C, et al. Electrochemical performance of Ti_3C_2 supercapacitors in KOH electrolyte. Journal of Advanced Ceramics. 2015;**4**:130-134. DOI: 10.1007/S40145-015-0143-3/METRICS

[47] Wang X, Mathis TS, Li K, Lin Z, Vlcek L, Torita T, et al. Influences from solvents on charge storage in titanium carbide MXenes. Nature Energy. 2019;**4**:241-248. DOI: 10.1038/s41560-019-0339-9

[48] Zhan C, Naguib M, Lukatskaya M, Kent PRC, Gogotsi Y, Jiang DE. Understanding the MXene pseudocapacitance. Journal of Physical Chemistry Letters. 2018;**9**:1223-1228. DOI: 10.1021/ACS.JPCLETT.8B00200

[49] Lin Z, Simon P. MXenes for supercapacitor application. In: 2D Metal Carbides and Nitrides (MXenes): Structure, Properties and Applications. Springer International Publishing; 2019. pp. 349-365. DOI: 10.1007/978-3-030-19026-2_18/COVER

[50] Simon P, Gogotsi Y, Dunn B. Where do batteries end and supercapacitors begin? Science. 2014;**343**:1210-1211. DOI: 10.1126/SCIENCE.1249625

[51] Okubo M, Hosono E, Kim J, Enomoto M, Kojima N, Kudo T, et al. Nanosize effect on high-rate Li-ion intercalation in $LiCoO_2$ electrode. Journal

of the American Chemical Society. 2007;**129**:7444-7452. DOI: 10.1021/JA0681927/ASSET/IMAGES/MEDIUM/JA0681927N00001.GIF

[52] Shimizu W, Makino S, Takahashi K, Imanishi N, Sugimoto W, Shimizu W, et al. Development of a 4.2 V aqueous hybrid electrochemical capacitor based on MnO_2 positive and protected Li negative electrodes. Journal of Power Sources. 2013;**241**:572-577. DOI: 10.1016/J.JPOWSOUR.2013.05.003

[53] Boisset A, Athouël L, Jacquemin J, Porion P, Brousse T, Anouti M. Comparative performances of birnessite and cryptomelane MnO_2 as electrode material in neutral aqueous lithium salt for supercapacitor application. Journal of Physical Chemistry C. 2013;**117**:7408-7422. DOI: 10.1021/JP3118488/ASSET/IMAGES/MEDIUM/JP-2012-118488_0016.GIF

[54] Thirumurugan A, Ramadoss A, Dhanabalan SS, Kamaraj SK, Chidhambaram N, Gobalakrishnan S, et al. MXene/ferrite magnetic nanocomposites for electrochemical supercapacitor applications. Micromachines. 2022;**13**:1792. DOI: 10.3390/MI13101792

[55] Cai Y, Shen J, Ge G, Zhang Y, Jin W, Huang W, et al. Stretchable $Ti_3C_2T_x$ MXene/carbon nanotube composite based strain sensor with ultrahigh sensitivity and tunable sensing range. ACS Nano. 2018;**12**:56-62. DOI: 10.1021/ACSNANO.7B06251

[56] Jiang C, Wu C, Li X, Yao Y, Lan L, Zhao F, et al. All-electrospun flexible triboelectric nanogenerator based on metallic MXene nanosheets. Nano Energy. 2019;**59**:268-276. DOI: 10.1016/J.NANOEN.2019.02.052

[57] Seyedin S, Zhang J, Usman KAS, Qin S, Glushenkov AM, Yanza ERS, et al. Facile solution processing of stable MXene dispersions towards conductive composite fibers. Global Challenges. 2019;**3**:1900037. DOI: 10.1002/GCH2.201900037

[58] Wang L, Chen L, Song P, Liang C, Lu Y, Qiu H, et al. Fabrication on the annealed $Ti_3C_2T_x$ MXene/epoxy nanocomposites for electromagnetic interference shielding application. Composites Part B: Engineering. 2019;**171**:111-118. DOI: 10.1016/J.COMPOSITESB.2019.04.050

[59] Zhao MQ, Torelli M, Ren CE, Ghidiu M, Ling Z, Anasori B, et al. 2D titanium carbide and transition metal oxides hybrid electrodes for Li-ion storage. Nano Energy. 2016;**30**:603-613. DOI: 10.1016/J.NANOEN.2016.10.062

[60] Tang W, He CE, Wang Y, Yang Y, Tsui CP. Low-temperature baroplastic processing of graphene-based polymer composites by pressure-induced flow. IOP Conference Series: Materials Science and Engineering. 2014;**62**:012023. DOI: 10.1088/1757-899X/62/1/012023

[61] Zheng W, Halim J, El Ghazaly A, Etman AS, Tseng EN, Persson POÅ, et al. Flexible free-standing MoO_3/$Ti_3C_2T_z$ MXene composite films with high gravimetric and volumetric capacities. Advanced Science. 2021;**8**:1-9. DOI: 10.1002/ADVS.202003656

[62] Xie W, Wang Y, Zhou J, Zhang M, Yu J, Zhu C, et al. MOF-derived $CoFe_2O_4$ nanorods anchored in MXene nanosheets for all pseudocapacitive flexible supercapacitors with superior energy storage. Applied Surface Science. 2020;**534**:147584. DOI: 10.1016/J.APSUSC.2020.147584

[63] Vahidmohammadi A, Moncada J, Chen H, Kayali E, Orangi J, Carrero CA, et al. Thick and freestanding MXene/

PANI pseudocapacitive electrodes with ultrahigh specific capacitance. Journal of Materials Chemistry A. 2018;**6**:22123-22133. DOI: 10.1039/C8TA05807E

[64] Boota M, Pasini M, Galeotti F, Porzio W, Zhao MQ, Halim J, et al. Interaction of polar and nonpolar polyfluorenes with layers of two-dimensional titanium carbide (MXene): Intercalation and pseudocapacitance. Chemistry of Materials. 2017;**29**:2731-2738. DOI: 10.1021/ACS.CHEMMATER.6B03933

[65] Su D, Zhang H, Zhang J, Zhao Y. Design and synthesis strategy of MXenes-based anode materials for sodium-ion batteries and progress of first-principles research. Molecules. 2023;**28**:6292. DOI: 10.3390/MOLECULES28176292

[66] Pu JH, Zhao X, Zha XJ, Bai L, Ke K, Bao RY, et al. Multilayer structured AgNW/WPU-MXene fiber strain sensors with ultrahigh sensitivity and a wide operating range for wearable monitoring and healthcare. Journal of Materials Chemistry A. 2019;**7**:15913-15923. DOI: 10.1039/C9TA04352G

[67] Li XP, Li Y, Li X, Song D, Min P, Hu C, et al. Highly sensitive, reliable and flexible piezoresistive pressure sensors featuring polyurethane sponge coated with MXene sheets. Journal of Colloid and Interface Science. 2019;**542**:54-62. DOI: 10.1016/J.JCIS.2019.01.123

[68] Wang F, Yang CH, Duan M, Tang Y, Zhu JF. TiO_2 nanoparticle modified organ-like Ti_3C_2 MXene nanocomposite encapsulating hemoglobin for a mediator-free biosensor with excellent performances. Biosensors & Bioelectronics. 2015;**74**:1022-1028. DOI: 10.1016/J.BIOS.2015.08.004

[69] Rakhi RB, Nayuk P, Xia C, Alshareef HN. Novel amperometric glucose biosensor based on MXene nanocomposite. Scientific Reports. 2016;**6**:1-10. DOI: 10.1038/srep36422

[70] Zhang L, Or SW. Self-assembled three-dimensional macroscopic graphene/MXene-based hydrogel as electrode for supercapacitor. APL Materials. 2020;**8**:91101. DOI: 10.1063/5.0015426/122961

[71] Venkateshalu S, Grace AN. $Ti_3C_2T_x$ MXene and vanadium nitride/porous carbon as electrodes for asymmetric supercapacitors. Electrochimica Acta. 2020;**341**:136035. DOI: 10.1016/J.ELECTACTA.2020.136035

[72] Sikdar A, Dutta P, Deb SK, Majumdar A, Padma N, Ghosh S, et al. Spontaneous three-dimensional self-assembly of MXene and graphene for impressive energy and rate performance pseudocapacitors. Electrochimica Acta. 2021;**391**:138959. DOI: 10.1016/J.ELECTACTA.2021.138959

[73] Wang K, Zheng B, Mackinder M, Baule N, Qiao H, Jin H, et al. Graphene wrapped MXene via plasma exfoliation for all-solid-state flexible supercapacitors. Energy Storage Materials. 2019;**20**:299-306. DOI: 10.1016/J.ENSM.2019.04.029

[74] Li X, Ma Y, Shen P, Zhang C, Cao M, Xiao S, et al. An ultrahigh energy density flexible asymmetric microsupercapacitor based on $Ti_3C_2T_x$ and PPy/MnO_2 with wide voltage window. Advanced Materials Technologies. 2020;**5**:2000272. DOI: 10.1002/ADMT.202000272

[75] Yu J, Zeng M, Zhou J, Chen H, Cong G, Liu H, et al. A one-pot synthesis of nitrogen doped porous MXene/TiO_2 heterogeneous film for high-performance flexible energy storage. Chemical Engineering Journal. 2021;**426**:130765. DOI: 10.1016/J.CEJ.2021.130765

[76] Liu Q, Yang J, Luo X, Miao Y, Zhang Y, Xu W, et al. Fabrication of a fibrous MnO_2@MXene/CNT electrode for high-performance flexible supercapacitor. Ceramics International. 2020;**46**:11874-11881. DOI: 10.1016/J.CERAMINT.2020.01.222

[77] Huang YL, Bian SW. Vacuum-filtration assisted layer-by-layer strategy to design MXene/carbon nanotube@ MnO_2 all-in-one supercapacitors. Journal of Materials Chemistry A. 2021;**9**:21347-21356. DOI: 10.1039/D1TA06089A

[78] Bao Z, Lu C, Cao X, Zhang P, Yang L, Zhang H, et al. Role of MXene surface terminations in electrochemical energy storage: A review. Chinese Chemical Letters. 2021;**32**:2648-2658. DOI: 10.1016/j.cclet.2021.02.012

[79] Akhter R, Maktedar SS. MXenes: A comprehensive review of synthesis, properties, and progress in supercapacitor applications. Journal of Materiomics. 2023;**9**:1196-1241. DOI: 10.1016/J.JMAT.2023.08.011

[80] Ghidiu M, Lukatskaya MR, Zhao M-Q, Gogotsi Y, Barsoum MW. Conductive two-dimensional titanium carbide 'clay' with high volumetric capacitance. Nature. 2014;**516**:78-81. DOI: 10.1038/nature13970

[81] Khazaei M, Mishra A, Venkataramanan NS, Singh AK, Yunoki S. Recent advances in MXenes: From fundamentals to applications. Current Opinion in Solid State & Materials Science. 2019;**23**:164-178. DOI: 10.1016/j.cossms.2019.01.002

[82] Althubiti NA, Aman S, Taha TAM. Synthesis of $MnFe_2O_4$/MXene/NF nanosized composite for supercapacitor application. Ceramics International. 2023;**49**:27496-27505. DOI: 10.1016/j.ceramint.2023.06.025

[83] Villers D, Jobin D, Soucy C, Cossement D, Chahine R, Breau L, et al. The influence of the range of electroactivity and capacitance of conducting polymers on the performance of carbon conducting polymer hybrid supercapacitor. Journal of the Electrochemical Society. 2003;**150**:A747. DOI: 10.1149/1.1571530/XML

[84] Laforgue A, Simon P, Fauvarque JF, Sarrau JF, Lailler P. Hybrid supercapacitors based on activated carbons and conducting polymers. Journal of the Electrochemical Society. 2001;**148**:A1130. DOI: 10.1149/1.1400742/XML

[85] Muzaffar A, Ahamed MB, Deshmukh K, Thirumalai J. A review on recent advances in hybrid supercapacitors: Design, fabrication and applications. Renewable and Sustainable Energy Reviews. 2019;**101**:123-145. DOI: 10.1016/J.RSER.2018.10.026

[86] Hu M, Zhang H, Hu T, Fan B, Wang X, Li Z. Emerging 2D MXenes for supercapacitors: Status, challenges and prospects. Chemical Society Reviews. 2020;**49**:6666-6693. DOI: 10.1039/D0CS00175A

[87] Zhou J, Yu J, Shi L, Wang Z, Liu H, Yang B, et al. A conductive and highly deformable all-pseudocapacitive composite paper as supercapacitor electrode with improved areal and volumetric capacitance. Small. 2018;**14**:1-9. DOI: 10.1002/SMLL.201803786

[88] Jiang Q, Kurra N, Alhabeb M, Gogotsi Y, Alshareef HN. All pseudocapacitive MXene-RuO_2 asymmetric supercapacitors. Advanced Energy Materials. 2018;**8**:1703043. DOI: 10.1002/AENM.201703043

[89] Xu S, Dall'Agnese Y, Wei G, Zhang C, Gogotsi Y, Han W.

Screen-printable microscale hybrid device based on MXene and layered double hydroxide electrodes for powering force sensors. Nano Energy. 2018;**50**:479-488. DOI: 10.1016/J.NANOEN.2018.05.064

[90] Zhao R, Wang M, Zhao D, Li H, Wang C, Yin L. Molecular-level heterostructures assembled from titanium carbide MXene and Ni-Co-Al layered double-hydroxide nanosheets for all-solid-state flexible asymmetric high-energy supercapacitors. ACS Energy Letters. 2018;**3**:132-140. DOI: 10.1021/ACSENERGYLETT.7B01063/SUPPL_FILE/NZ7B01063_SI_001.PDF

[91] Pan Z, Cao F, Hu X, Ji X. A facile method for synthesizing CuS decorated Ti_3C_2 MXene with enhanced performance for asymmetric supercapacitors. Journal of Materials Chemistry A. 2019;**7**:8984-8992. DOI: 10.1039/C9TA00085B

[92] Chen Y, Kang Y, Zhao Y, Wang L, Liu J, Li Y, et al. A review of lithium-ion battery safety concerns: The issues, strategies, and testing standards, Journal of Energy. Chemistry. 2021;**59**:83-99. DOI: 10.1016/J.JECHEM.2020.10.017

[93] Dunn B, Kamath H, Tarascon JM. Electrical energy storage for the grid: A battery of choices. Science. 2011;**334**:928-935. DOI: 10.1126/SCIENCE.1212741/SUPPL_FILE/DUNN-SOM.PDF

[94] Tarascon JM, Armand M. Issues and challenges facing rechargeable lithium batteries. Nature. 2001;**414**:359-367. DOI: 10.1038/35104644

[95] Aricò AS, Bruce P, Scrosati B, Tarascon JM, Van Schalkwijk W. Nanostructured materials for advanced energy conversion and storage devices. Nature Materials. 2005;**4**:366-377. DOI: 10.1038/nmat1368

[96] Wang Y, Song Y, Xia Y. Electrochemical capacitors: Mechanism, materials, systems, characterization and applications. Chemical Society Reviews. 2016;**45**:5925-5950. DOI: 10.1039/C5CS00580A

[97] Han H, Cho S. Ex situ fabrication of polypyrrole-coated core-shell nanoparticles for high-performance coin cell supercapacitor. Nanomaterials. 2018;**8**:726. DOI: 10.3390/NANO8090726

[98] Simon P, Gogotsi Y. Materials for electrochemical capacitors. Nature Materials. 2008;**7**:845-854. DOI: 10.1038/nmat2297

[99] Shen B, Hao R, Huang Y, Guo Z, Zhu X. Research progress on MXene-based flexible supercapacitors: A review. Crystals. 2022;**12**:1099. DOI: 10.3390/CRYST12081099

[100] Naguib M, Mochalin VN, Barsoum MW, Gogotsi Y. 25th anniversary article: MXenes: A new family of two-dimensional materials. Advanced Materials. 2014;**26**:992-1005. DOI: 10.1002/adma.201304138

[101] Yi Q, Pei X, Das P, Qin H, Lee SW, Esfandyarpour R. A self-powered triboelectric MXene-based 3D-printed wearable physiological biosignal sensing system for on-demand, wireless, and real-time health monitoring. Nano Energy. 2022;**101**:107511. DOI: 10.1016/J.NANOEN.2022.107511

[102] Wang Y, Wang Y. Recent progress in MXene layers materials for supercapacitors: High-performance electrodes. SmartMat. 2023;**4**:1-35. DOI: 10.1002/smm2.1130

[103] Bhat A, Anwer S, Bhat KS, Mohideen MIH, Liao K, Qurashi A. Prospects challenges and stability of 2D MXenes for clean energy conversion

and storage applications. npj 2D Materials and Applications. 2021;**5**:61. DOI: 10.1038/s41699-021-00239-8

[104] Jaya Prakash N, Kandasubramanian B. Nanocomposites of MXene for industrial applications. Journal of Alloys and Compounds. 2021;**862**:158547. DOI: 10.1016/J.JALLCOM.2020.158547

Chapter 5

Thermal Performance Analysis of RT 27 as PCM in Double-Pipe Heat Exchangers: Concentric and Hairpin Heat Exchangers

Pallavi Kumari and Debasree Ghosh

Abstract

To reduce the usage of conventional fuel and lessen environmental degradation, the utilization of phase change materials presents a promising revenue. By comparing the performance of phase change material melting and solidification of hairpin and concentric-type heat exchangers, the study aims to design latent heat energy storage systems for domestic and industrial applications. The study further reveals that the melting/solidification time depends on various parameters including thermal diffusivity and viscosity of the phase change material (PCM). Furthermore, the energy stored in hairpin and concentric is the same for both the cases as the amount of PCM taken is the same for both the cases but the time required for storing energy is less in case of hairpin, that is, 189 min than concentric, that is, 318.06 min. Additionally, the study emphasizes the importance of PCM selection, highlighting that PCMs with higher latent heat values can store more energy, but the rate of energy storage depends on the temperature difference between the high-temperature fluid and the initial temperature of the PCM. Overall, the findings suggest that through effective numerical analysis and optimization of design parameters, it is possible to propose energy storage systems that maximize efficiency and capacity, thereby contributing to the reduction of conventional fuel consumption and environmental impact.

Keywords: concentric double-pipe heat exchanger, hairpin double-pipe heat exchanger, latent heat thermal energy storage, enthalpy-porosity model, melting/solidification

1. Introduction

During the last four decades, latent heat thermal energy storage has become a vital topic, and researchers are still working to find the best energy storage technique for various materials [1]. Using phase change material (PCM) becomes the most suitable option for this purpose as it shows isothermal behavior during the phase change with high thermal storage capacity per unit volume [2, 3]. The performance of the latent heat thermal energy storage (LHTES) mainly depends on the thermophysical properties of the PCM and the design of heat exchangers [4]. Hosseini et al. [5] performed experimental and numerical research on shell and tube heat exchangers using RT 50 paraffin wax. They studied heat transfer specification and thermal behavior during the melting and solidification process [6]. Esapour et al. studied the melting process of a multi-tube heat exchanger (HEX) using RT35 as PCM. The results showed that the melting time is reduced by 29% by increasing the number of tubes on the shell side [7]. Ismail et al. experimentally studied the solidification of PCM, and the result showed that solidification time decreases when the fluid inlet temperature is lowered [8]. Due to the low thermal conductivity of PCM, the melting/solidification is prolonged and has limitations for the energy storage device. To solve this problem, various literature surveys have been done using multiple tubes in heat exchangers [9], external fins in multi-tube heat exchangers [10], microencapsulation of PCM [11] vertical shell, and tube heat exchangers [12]. Soliman et al. conducted experimental research on the thermal efficiency of sliding windows that incorporated single PCM (RT 27 or RT 35) and multiple PCMs (both RT 27 and RT 35) combined with PV cells. Their findings indicated that utilizing multiple PCMs could lower the maximum interior wall temperature by up to 46%, suggesting that such innovative designs could be beneficial for building applications [13]. Nagaraju et al. performed a numerical study to design a helmet intended to provide cooling for the face and head using two types of PCMs: RT 27 and capric acid. Their results indicated that RT 27 reached the maximum liquid fraction in 150 minutes compared to 175 minutes for capric acid when the system was positioned horizontally [14]. Another study conducted a numerical investigation of PCM (paraffins C18 and RT 27) in a finned rectangular PCM-air heat exchanger. This study examined the impact of fin quantity on heat transfer enhancement and the effect of natural convection on the solidification time of PCM. The results showed that after 1 hour and 30 minutes, 55% of RT 27 was solidified compared to 40% of C18 paraffin, highlighting the significant improvement in heat transfer during the discharging phase with RT 27 [15]. Faizan et al. studied the numerical influence of different gravitational conditions (earth, moon, and micro-gravity) on the melting behavior of RT 27 PCM. They found that the PCM melt time was twice as long for moon gravity and 5–6 times longer for microgravity compared to earth gravity [16]. Kumirai et al. experimentally examined the phase change behavior of three commercially available PCMs: two paraffinic PCMs (RT 27 and RT 25) and one salt hydrate PCM. Their study found that paraffinic PCMs absorbed heat more quickly than salt hydrate PCM [17]. Further, the authors compared the energy capacity of concentric and hairpin heat exchangers using RT 50 as PCM. They found that the energy capacity of the hairpin HEX was 3.25 times higher than that of the concentric HEX [18]. They also conducted a numerical study on the melting and solidification processes of a U-tube heat exchanger using three different PCMs (RT50, RT 35, and RT 27) to assess the effects of thermophysical properties and dimensionless numbers related to heat transfer [19]. Despite reviewing previous studies, no research was found that utilized a U-tube type heat exchanger with RT 27 as a PCM. In this study, we compare the melting and solidification processes for both U-tube and concentric heat exchangers using RT 27 paraffin wax as the PCM.

2. Computational domain

In this study, we examine two double-pipe heat exchangers (HEX) to investigate the phase change process, that is, melting/solidification. The overall equipment's geometry is illustrated in **Figures 1** and **2**. **Figure 1** depicts the setup with two concentric pipes, where the high/low temperature fluid (HTF/LTF) flows through the inner tube. Also, it shows an alternative configuration where the inner tube is U-shaped, forming a hairpin arrangement within the pipe containing the phase change material (PCM). The outer pipe, which contains paraffin wax as the PCM, is sealed at both ends to prevent the escape of the PCM, ensuring efficient latent heat energy storage. However, practical designs should include provisions for the expansion of PCM. For both the HEX, the inner and outer diameter (ID/OD) of outer pipe are 140 and 144 mm and dimensions of inner pipe are 70 mm (ID), 72 mm (OD) and 20 mm (ID), 24 mm (OD) for concentric and hairpin HEX respectively. The total length of the inner pipe is taken as 900 and 1600 mm for CHEX and HHEX, respectively [18]. The thermophysical properties of the PCM are listed in **Table 1**.

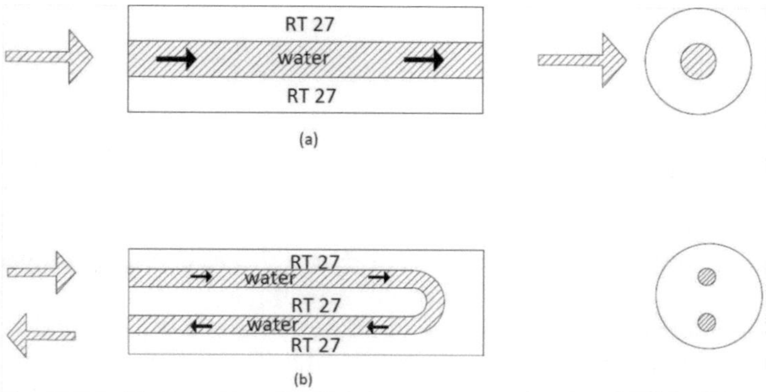

Figure 1.
Schematic diagram of the computational domain.

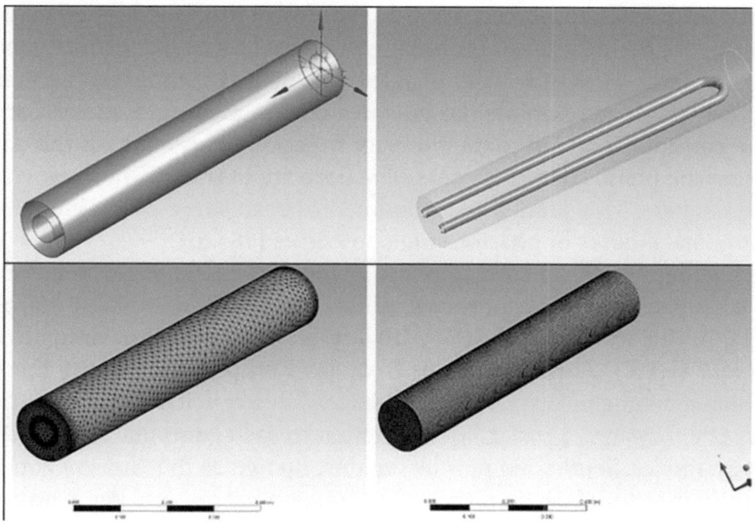

Figure 2.
Computational domain and meshing [18].

Name	Specifications
Pcm	RT 27
Solidus Temperature *(K)*	301
Liquidus Temperature *(K)*	303
Specific Heat Capacity *(J/kg K)*	2000
Latent Heat *(kJ/kg)*	179
Density *(kg/m*^{3}*)*	815
Thermal Conductivity *(W/m K)*	0.15
Viscosity *(kg/m s)*	0.00342
Thermal Expansion Coefficient *(*K^{-1}*)*	0.0005

Table 1.
Specification of PCM [19].

3. Mathematical model

The phase change process involves the transformation of a solid phase into a liquid phase, with distinct flow and heat transfer mechanisms in each phase. In the solid phase, heat transfer occurs primarily through conduction, whereas in the liquid phase, both conduction and convection are present. Natural convection or convection in the liquid phase arises due to density differences. The liquid phase demonstrates Newtonian flow characteristics, and when a solid phase coexists within a solid-liquid mixture, the system behaves similarly to porous media. The enthalpy-porosity model accurately describes the flow mechanisms of both phases in this mixture. The porosity of the mixture, determined by the liquid fraction or melt fraction, ranges from zero in the solid phase to one in the liquid phase. To account for the flow effects with changing porosity, a source term is introduced into the momentum balance equation. This velocity-proportional source term uses a porosity function to gradually reduce velocities in the phase-changing computational components from finite values in the liquid phase to zero in the fully solid phase. By using this technique, the momentum equation can be made to resemble the flow in a porous medium Kozeny-Carman eq. A mushy zone constant is added to the porosity function to account for the dampening impact during the phase change process. The porosity of the PCM varies with the melt or liquid percentage. This modeling technique has been successfully implemented in various numerical studies of phase change processes [18–20].

Various methods exist for solving phase change processes, including the enthalpy method, heat capacity method, porous medium method, and enthalpy-porosity method [21]. In this study, we use the enthalpy-porosity model to simulate the PCM heat exchanger (HEX). This approach is frequently employed to create numerical models of phase change materials (PCMs). Instead of explicitly calculating the melting front, the method assigns a scalar liquid fraction to each computational node within a single domain model. Semi-solid porous regions, known as the "mushy zone," form at the solid/liquid interface based on the liquid fraction value. Two key parameters for this method are the mushy zone constant and the phase change temperature interval. While the impact of the mushy zone constant has been thoroughly researched for various PCMs, its relationship with the phase transition temperature range is less understood. This method calculates the total energy (h) of a material as the sum of its latent enthalpy (hl) and sensible enthalpy (hs). In the study, the PCM heat exchanger uses water as the heat transfer fluid (HTF) and paraffin wax as the PCM. During the heat transfer process, water stays in the liquid state, while the PCM transitions from solid to liquid. The enthalpy-porosity model's advantage is that the governing equations, discussed below, apply to both water and the solid and liquid phases of the PCM. The HTF flow through the heat exchanger is assumed to be laminar. Both conduction and convection in the PCM are considered in this study. Additionally, the thermophysical properties, such as paraffin's density, thermal conductivity, and viscosity, are assumed to be constant during the transient process.

The buoyancy effects in the PCM are modeled using the Boussinesq approximation, and the PCM is considered homogeneous and isotropic.

3.1 Governing equations are as follows

Mass conservation equation [19]:

$$\nabla.\left(\rho\vec{v}\right) = 0 \tag{1}$$

Momentum balance equation:

$$\frac{\partial\left(\rho\vec{v}\right)}{\partial t} + \nabla.\left(\rho\vec{v}\vec{v}\right) = -\nabla \mathrm{P} + \nabla.\left(\mu\nabla\vec{v}\right) + \rho\vec{g}\beta(T - T_{ref}) + \vec{S} \tag{2}$$

where,

$$\vec{S} = -A(\gamma)\vec{v} = -C_{mushy}\frac{(1-\gamma)^2}{\gamma^3 + 0.001}\vec{v} \tag{3}$$

$$\gamma = \begin{cases} 0, T < T_s \\ \dfrac{T - T_s}{T_l - T_s}, T_s < T < T_l \\ 1, T > T_l \end{cases} \tag{4}$$

The equation for energy balance:

$$\frac{\partial(\rho H)}{\partial t} + \nabla.\left(\rho\vec{v}H\right) = \nabla.(\mathrm{k}\nabla T) \tag{5}$$

The overall enthalpy is described as

$$H = h + \Delta H = \left(h_{ref} + \int_{T_{ref}}^{T} C_p dT\right) + \gamma L \tag{6}$$

3.2 Initial and boundary conditions

In this study, paraffin wax, RT27, is used as a phase change material (PCM). The melting (charging) processes of the PCM are examined, beginning with an initial temperature set one degree below each PCM's solidus temperature and one degree above the PCM's liquidus temperature during solidification (discharging). To ensure a consistent Stefan number across all experiments, the temperature of the high-temperature fluid (HTF) at the boundary is adjusted accordingly. The research is conducted with Stefan's numbers $St_{PCM} = \left|\frac{Cp_{PCM}(T_{m,PCM}-T_w)}{L_{PCM}}\right|$: 0.23. The initial temperature for HTF and LTF are considered as 323 K and 281 K, respectively. Thus, this setup is also referred to as a low-temperature heat source. In this simulation, we assume the thermophysical properties remain constant, with density variations modeled using the Boussinesq approximation. The heat transfer process is simulated using ANSYS Fluent.

4. Numerical scheme and validation

For the hairpin double-pipe HEX, 98,281 triangular cells and 196,311 cells are used in the phase change process analysis of the double-pipe heat exchanger (HEX). The three-dimensional solid model and mesh are depicted in **Figure 2**. The enthalpy-porosity model is applied for discretization of governing equations. The SIMPLE algorithm is opted with a pressure-based solver for this study. Also, to discretize momentum and energy equations, a second-order upwind scheme is applied, followed by the PRESTO scheme for pressure correction. Ghosh et al. established grid independence for the concentric double-pipe HEX [20]. The grid independence test for the hairpin HEX is studied using three grids where Grid 1, Grid 2, and Grid 3 refer to 603,908 cells, 655,056 cells, and 924,194 cells, respectively. The optimized time step taken for simulation is 0.1 s. As Grid 1 shows a larger deviation, the simulation was carried out further using Grid 2 to minimize computational time for HEXs. Using the enthalpy-porosity technique in ANSYS Fluent, the study imitated the work of Gorzin et al. with comparable results in order to confirm the numerical model [20, 22].

5. Results and discussion

Melting and solidification are vital phase change processes in thermal management systems, especially in heat exchangers utilized for energy storage, heating, cooling, and material processing. These processes involve a substance transitioning between solid and liquid phases, accompanied by the absorption or release of latent heat. A thorough understanding of these phase change processes is crucial for designing efficient heat exchangers, optimizing energy consumption, and improving overall system performance.

5.1 Influence of heat exchanger type on charging or melting of PCM

The heat exchanger is an essential device for recovering heat from waste fluids before they are released into the environment and for using this recovered heat to preheat process fluids. The Number of Transfer Units (NTU) method has shown that a heat exchanger's effectiveness is maximized during the phase change of one of the fluids. In this study, two double-pipe heat exchangers (HEXs) with identical shell diameters and lengths were analyzed: one concentric double-pipe HEX and one hairpin double-pipe HEX. Paraffin wax (RT 27) served as the phase change material

(PCM) with a hot thermal fluid (HTF) flow rate of 0.05 kg/s, and an inlet temperature of 323 K. **Figure 3** illustrates the complete melting process over time for both HEXs. The masses of paraffin wax used were 10.735 kg for the hairpin HEX and 8.3 kg for the concentric HEX. The time required for complete melting of the PCM was 189 minutes for the hairpin HEX and 331 minutes for the concentric HEX, indicating a longer melting time for the concentric HEX. The melting process can be analyzed by observing the progression of the melting front, which can be divided into four distinct phases. During the initial phase, the melting front moves outward parallel to the wall of the HTF tube, primarily influenced by viscous forces. At the start, the melting rate is at its highest but then decreases significantly. This phenomenon can be explained by the initially high-temperature difference, which gradually diminishes as melting progresses. **Figure 4** shows that the temperature difference between the inlet and outlet of the HTF is greater for the hairpin HEX compared to the concentric HEX, suggesting that the hairpin HEX extracts more heat from the HTF. This implies that reducing the inlet area or diameter of the U-tube could potentially increase the heat storage rate. The heat transfer area between the PCM and HTF is consistent for both HEX types.

However, the reduced inlet area increases the velocity and Reynolds number of the HTF, thereby accelerating the charging process. The contours (**Figure 5**) show that the liquid PCM is visible at around 5 min but slightly delayed in case of concentric

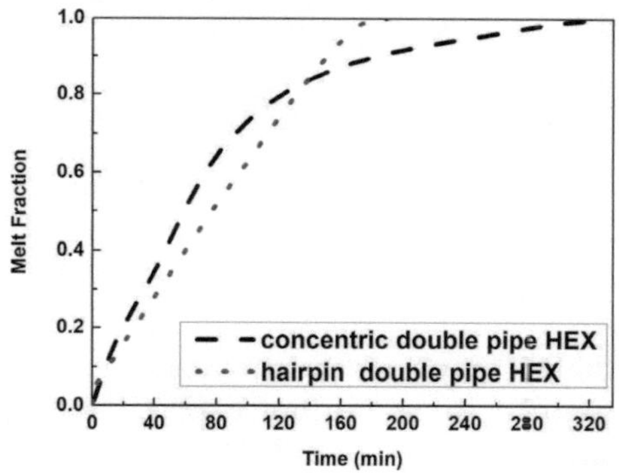

Figure 3.
Influence of melt fraction during melting.

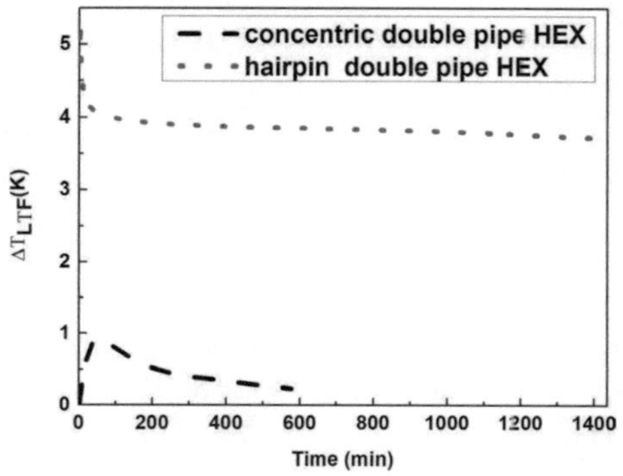

Figure 4.
Influence of inlet-outlet temperature difference of HTF during melting.

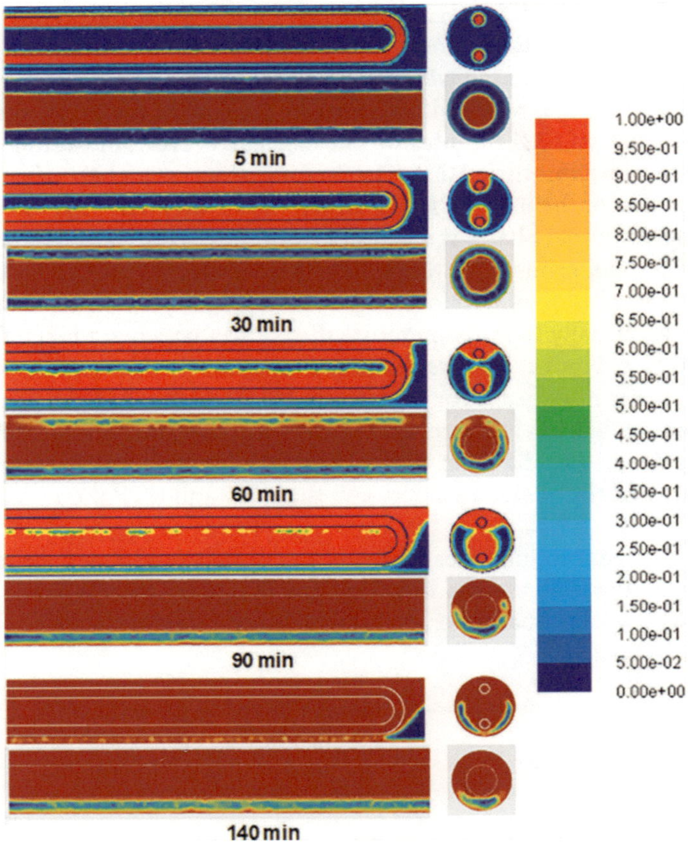

Figure 5.
Contour of melt-fraction concentric and hairpin double-pipe heat exchanger (St = 0.23).

HEX. It means that liquid PCM is observed where the melting process is controlled by natural convection. In contrast, in the concentric HEX, lower inlet velocity and Reynolds number delay the appearance of liquid PCM, with heat transfer dominated by conduction. Clearly, we can say that the rate of melting is higher in case of hairpin HEX, and after 30 min, the natural convection is dominating for both cases. Due to the formation of the dead zone at the bottom right-hand side of the hairpin HEX (140 min), the process slows down as the distance between the inner pipe and PCM increases. In this zone, conduction dominates as there is no natural convection visible, and thermal conductivity decreases, which slows the process. The energy stored in the concentric HEX is 2167 kJ, while in the hairpin HEX it is 2164 kJ, indicating nearly identical energy storage. However, the time required to store this energy in the hairpin HEX is 189 minutes, significantly less than the 318 minutes required for the concentric HEX. Observations show that the heat transfer area at the inlet of the hairpin HEX is smaller than that of the concentric HEX. Since area is inversely proportional to velocity, this results in the HTF flowing with a higher Reynolds number through the inner pipe of the hairpin HEX. This work shows that while the energy extracted from HTF is determined by the initial temperature of the HTF and the PCM, the quantity of energy stored in PCM is dependent on both latent heat and the rate of energy storage. Thus, the melting time is determined by the temperature differential between the HTF and the initial PCM temperature.

5.2 Influence of heat exchanger type on discharging or solidification of PCM

The simulated graph in **Figure 6** illustrates the complete solidification process over time for both HEXs. The graph shows that the time required to solidify paraffin wax (RT 27) using the hairpin HEX (approximately 1435 minutes) is significantly longer than using the concentric HEX (approximately 577 minutes). At the start of the discharging process, natural convection within the liquid PCM is the primary mode of heat transfer within the storage unit. This convection efficiently transfers heat, quickly removing the PCM's sensible heat. Sensible heat refers to the energy that causes a temperature change without altering the material's phase. In this scenario, the PCM's sensible heat decreases rapidly, leading to a quick temperature drop until the freezing point is reached. As the PCM adjacent to the HTF tube continues to lose heat and cools down, it eventually solidifies. During this phase transition from liquid to solid, the PCM's thermal properties undergo significant changes. In its solid form, the PCM has much lower thermal conductivity compared to its liquid state. Thermal conductivity, which measures a material's ability to conduct heat, is crucial here. Consequently, the rate of heat transfer from the HTF to the PCM diminishes considerably as the PCM solidifies around the HTF tube. This reduction occurs because the insulating nature of the solid PCM hinders the flow of heat from the HTF to the PCM. **Figures 7** and **8** show the time-dependent variation in heat flux for RT 27 during the melting and solidification process for both HEXs, all having the same Stefan number

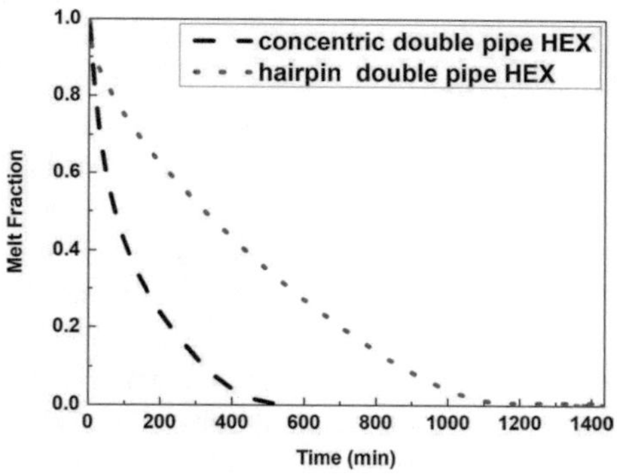

Figure 6.
Influence of melt fraction during solidification.

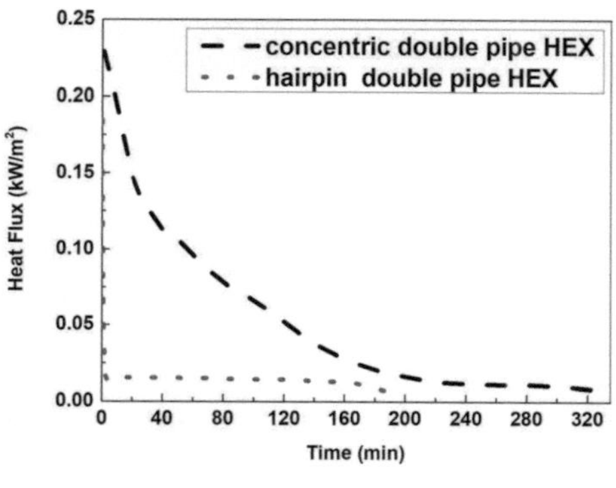

Figure 7.
Influence of heat flux of PCM during melting.

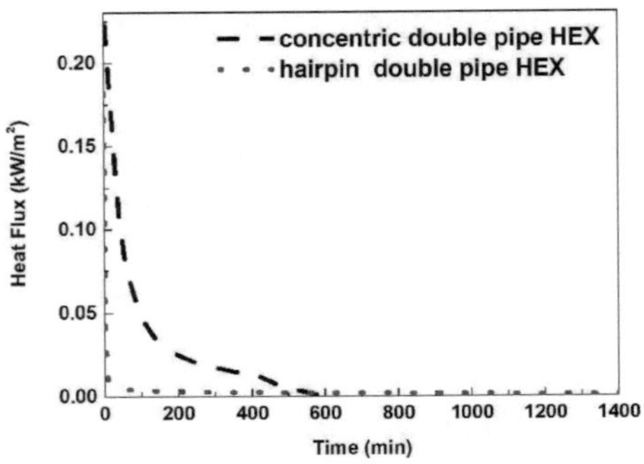

Figure 8.
Influence of heat flux of PCM during solidification.

(St = 0.23). The heat flux decreases over time in both processes. Initially, there is a substantial rate of change in heat flux, which gradually diminishes as the melting or solidification progresses. Notably, following the formation of a melt fraction, the heat flux approaches a near-minimum level, and the rate of change in heat flux significantly slows, as illustrated in **Figures** 7 and **8**.

During the charging process, the heat flux ranges from 0.18 to 0.006 kW/m^2 for the hairpin HEX and from 0.22 to 0.004 kW/m^2 for the concentric HEX. During the solidification process, the heat flux ranges from 0.18 to 0.00033 kW/m^2 for the hairpin HEX and from 0.22 to 0.00064 kW/m^2 for the concentric HEX. This indicates that at the start of the process, the contact time between the PCM and HTF is at its maximum, while the heat transfer rate is relatively low. In contrast, for the HHEX, the contact time is notably low due to the high velocity of the HTF. Conversely, in the concentric HEX, the extended contact duration between the HTF and the PCM facilitates more efficient heat transfer. This prolonged interaction allows the PCM to absorb heat from the HTF over a more extended timeframe. Consequently, the velocity of the HTF decreases in the concentric HEX, leading to an increase in the melting rate. In the hairpin HEX, the length of the pipe is nearly double that of the concentric double-pipe HEX due to the use of a U-shaped tube. This configuration results in higher HTF velocities, thereby enhancing the charging process and enabling higher energy storage in a shorter time from the same heat source (**Figure 9**).

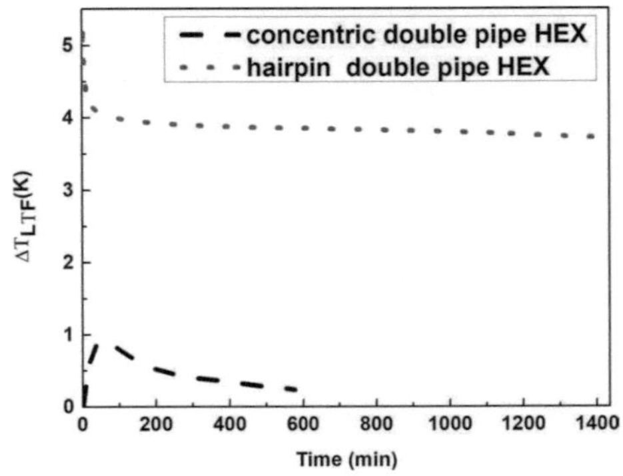

Figure 9.
Influence of inlet-outlet temperature difference of LTF during solidification.

The use of phase change materials (PCMs) represents a promising technology for thermal energy storage, offering applications across several domains:

Energy Storage Systems: Heat exchangers that efficiently handle melting and solidification are essential for thermal energy storage and vital for renewable energy systems like solar power. These systems store excess thermal energy during peak production periods and release it during low production or high demand times. For instance, solar power plants utilize latent heat thermal energy storage (LHTES) to store solar energy during the day and release it at night, ensuring a stable power supply.

HVAC Systems: PCMs in heat exchangers are employed in heating, ventilation, and air conditioning (HVAC) systems to boost energy efficiency and reduce operational costs. Buildings can use PCMs to store cool energy at night when energy costs are lower and release it during the day to lessen cooling loads.

Industrial Cooling: Efficient phase change in heat exchangers is crucial for industries requiring precise temperature control, such as food processing, pharmaceuticals, and chemical manufacturing. For example, pharmaceutical manufacturing employs heat exchangers with PCMs to maintain critical temperatures during storage and transportation.

Waste Heat Recovery: Systems designed to capture and reuse waste heat from industrial processes can utilize heat exchangers with PCMs to enhance efficiency and reduce energy consumption. Industrial plants implement these systems to capture and store waste heat, which can later be used to preheat raw materials or generate power.

However, using organic PCMs like RT 27 has limitations. During the summer season, the sensible heat transfer is very high as compared to latent heat, which impacts their effectiveness.

6. Conclusions

The study affects the heat exchanger design and concludes that the volume of the PCM can be increased by changing the design to hairpin type HEX to store more energy. Also, due to higher velocity of HTF in hairpin HEX the amount of energy stored is comparatively higher than concentric HEX. The efficiency of latent heat energy storage device is enhanced by optimizing a balance between energy storage rate and energy stored. Also, the study aid in selecting appropriate design with a PCM. Further, the study can be carried out by using various other PCM and a comparative study can be done for more accurate results.

To ensure the sustainable development of thermal energy management, further advancement of phase change material (PCM) technology is essential. Future studies should focus on several key aspects:

a. *Avoidance of PCM Phase Separation*: Future research should aim to maintain the stable performance of energy storage systems by preventing phase separation in PCMs.

b. *Environmental and Health Impact Assessments*: It is crucial to conduct assessments to evaluate the environmental and health impacts of widespread PCM applications across various fields, ensuring their suitability and identifying any potential risks.

c. *Advanced Encapsulation Methods*: Developing advanced encapsulation techniques can improve the durability, reliability, and thermal performance of

PCMs, thereby ensuring their long-term effectiveness in real-world applications.

d. *Enhancing Thermal Conductivity and Heat Transfer Efficiency*: Exploring novel methods to enhance thermal conductivity and heat transfer efficiency of PCMs will maximize their energy storage and release capabilities.

e. *Evaluation of Fins, Metal Foam, and Nano-sized Particles*: Research should investigate the benefits of using fins, metal foam, or nano-sized particles to enhance thermal performance while maintaining the same volume constraints of energy storage systems, thus maximizing efficiency.

f. *Standardized PCM Selection Criteria*: Establishing clear and uniform criteria for selecting PCMs is necessary to ensure the optimal material is chosen for each specific application.

These areas of focus will drive the advancement of PCM technology, contributing to more efficient and sustainable thermal energy management solutions.

Acknowledgements

The authors acknowledge the financial support provided by the Science and Engineering Research Board (SERB) of India under grant number SPG/2021/004530. This funding was crucial in supporting our research efforts described in this work.

Conflict of interest statement

The authors declare that they have no conflicts of interest to disclose.

Author details

Pallavi Kumari and Debasree Ghosh*
Department of Chemical Engineering, Birla Institute of Technology, Mesra, Ranchi, India

*Address all correspondence to: dghosh@bitmesra.ac.in

References

[1] Choure BK, Alam T, Kumar R. A review on heat transfer enhancement techniques for PCM based thermal energy storage system. Journal of Energy Storage. 2023;**72**:108161. DOI: 10.1016/j.est.2023.108161

[2] Joulin A, Younsi Z, Zalewski L, Lassue S, Rousse DR, Cavrot J-P. Experimental and numerical investigation of a phase change material: Thermal-energy storage and release. Applied Energy. 2011;**88**(7):2454-2462. DOI: 10.1016/j.apenergy.2011.01.036

[3] Medrano M, Yilmaz MO, Nogués M, Martorell I, Roca J, Cabeza LF. Experimental evaluation of commercial heat exchangers for use as PCM thermal storage systems. Applied Energy. 2009; **86**(10):2047-2055. DOI: 10.1016/j.apenergy.2009.01.014

[4] Kalapala L, Devanuri JK. Influence of operational and design parameters on the performance of a PCM based heat exchanger for thermal energy storage – A review. Journal of Energy Storage. 2018;**20**:497-519. DOI: 10.1016/j.est.2018.10.024

[5] Hosseini MJ, Ranjbar AA, Sedighi K, Rahimi M. A combined experimental and computational study on the melting behavior of a medium temperature phase change storage material inside shell and tube heat exchanger. International Communications in Heat and Mass Transfer. 2012;**39**(9): 1416-1424. DOI: 10.1016/j.icheatmasstransfer.2012.07.028

[6] Hosseini MJ, Rahimi M, Bahrampoury R. Experimental and computational evolution of a shell and tube heat exchanger as a PCM thermal storage system. International Communications in Heat and Mass Transfer. 2014;**50**:128-136. DOI: 10.1016/j.icheatmasstransfer.2013.11.008

[7] Esapour M, Hosseini MJ, Ranjbar AA, Pahamli Y, Bahrampoury R. Phase change in multi-tube heat exchangers. Renewable Energy. 2016;**85**:1017-1025. DOI: 10.1016/j.renene.2015.07.063

[8] Ismail KAR, Lino FAM, da Silva RCR, de Jesus AB, Paixão LC. Experimentally validated two dimensional numerical model for the solidification of PCM along a horizontal long tube. International Journal of Thermal Sciences. 2014;**75**:184-193. DOI: 10.1016/j.ijthermalsci.2013.08.008

[9] Agyenim F, Eames P, Smyth M. Heat transfer enhancement in medium temperature thermal energy storage system using a multitube heat transfer array. Renewable Energy. 2010;**35**(1): 198-207. DOI: 10.1016/j.renene.2009.03.010

[10] Sayehvand H-O, Abolfathi S, Keshavarzian B. Investigating heat transfer enhancement for PCM melting in a novel multi-tube heat exchanger with external fins. Journal of Energy Storage. 2023;**72**:108702. DOI: 10.1016/j.est.2023.108702

[11] Kalaiselvam S, Veerappan M, Arul Aaron A, Iniyan S. Experimental and analytical investigation of solidification and melting characteristics of PCMs inside cylindrical encapsulation. International Journal of Thermal Sciences. 2008;**47**(7):858-874. DOI: 10.1016/j.ijthermalsci.2007.07.003

[12] Nie C, Liu J, Deng S. Effect of geometric parameter and nanoparticles on PCM melting in a vertical shell-tube system. Applied Thermal Engineering.

2021;**184**:116290. DOI: 10.1016/j.applthermaleng.2020.116290

[13] Soliman AS, Radwan A, Fouda MS, Sultan AA, Abdelrehim O. Energy assessment of a sliding window integrated with PV cell and multiple PCMs. Journal of Energy Storage. 2024; **86**:111341. DOI: 10.1016/j.est.2024.111341

[14] Nagaraju D, Santhosi BVSRN, Mohammad AR, Syed J, Kolla NK. Numerical investigation of sustainable thermal energy storage (TES) system for personal helmet cooling. International Journal of Thermofluids. 2023;**20**: 100481. DOI: 10.1016/j.ijft.2023.100481

[15] Jmal I, Baccar M. Numerical investigation of PCM solidification in a finned rectangular heat exchanger including natural convection. International Journal of Heat and Mass Transfer. 2018;**127**:714-727. DOI: 10.1016/j.ijheatmasstransfer.2018.08.058

[16] Faizan M, Alkaabi AK, Afgan I. Influence of variable gravity on the phase change material for enclosed vertical channels with helical fins. Applied Thermal Engineering. 2024;**250**: 123494. DOI: 10.1016/j.applthermaleng.2024.123494

[17] Kumirai T, Dirker J, Meyer J. Experimental analysis for thermal storage performance of three types of plate encapsulated phase change materials in air heat exchangers for ventilation applications. Journal of Building Engineering. 2019;**22**:75-89. DOI: 10.1016/j.jobe.2018.11.016

[18] Kumari P, Ghosh D. A comparative numerical analysis of concentric and hairpin heat exchanger for efficient energy storage using phase-change material. Journal of Thermal Analysis and Calorimetry. 2023;**148**(21): 12211-12224. DOI: 10.1007/s10973-023-12501-w

[19] Kumari P, Raj A, Ghosh D. Selection of phase change material for latent heat thermal energy storage using a hair-pin heat exchanger: Numerical study. Journal of Thermal Science and Engineering Applications. 2024;**16**(9): 1-33. DOI: 10.1115/1.4065490

[20] Ghosh D, Kumar P, Sharma S, Guha C, Ghose J. Numerical investigation on latent heat thermal energy storage in a phase change material using a heat exchanger. Heat Transfer. 2021;**50**(5):4289-4308. DOI: 10.1002/htj.22075

[21] Moreira M, Silva T, Dias-de-Oliveira J, Neto F, Amaral C. Numerical modelling of radiant systems and phase change materials in building applications - A review. Applied Thermal Engineering. 2023;**234**:121342. DOI: 10.1016/j.applthermaleng.2023.121342

[22] Gorzin M, Hosseini MJ, Ranjbar AA, Bahrampoury R. Investigation of PCM charging for the energy saving of domestic hot water system. Applied Thermal Engineering. 2018;**137**:659-668. DOI: 10.1016/j.applthermaleng.2018.04.016

Chapter 6

Revolutionizing Energy Applications: The Power of Interconnected Pores in Hierarchically Porous Carbon

Ananya Pal and Atanu Roy

Abstract

Electrode materials are crucial in electrochemical energy storage devices, prompting extensive research into optimizing high-performance options. Hierarchically porous carbons (HPCs), featuring 1D to 3D networks, are of great interest due to their excellent electrical conductivity, high surface area, unique physicochemical properties, and superior chemical stability. These properties make micro-/nanostructured porous carbon a promising candidate for energy storage technology. This chapter summarizes the design and synthesis of HPC materials using hard-templating, soft-templating, and nontemplating routes, with a focus on nontemplating strategies for biopolymers. It discusses recent use of HPCs and their composites in various electrochemical energy storage applications, such as supercapacitors, lithium-ion batteries, sodium-ion batteries, post-lithium-ion batteries, and hybrid energy storage devices. Moreover, the chapter offers insights into future challenges and research opportunities in HPC materials.

Keywords: hierarchically porous carbons, supercapacitors, hybrid energy storage device, lithium-ion batteries, post-lithium-ion batteries

1. Introduction

The rapidly changing global economy has led to a substantial rise in energy demands, raising serious concerns about energy shortages and the impact of global warming on society. Consequently, the need for highly efficient, clean, and renewable energy sources, such as wind, solar, and ocean energy, has become more critical and urgent than ever. The intermittent nature of these renewable resources makes energy storage an essential part of the convenient energy solution. To meet the global energy demand, current state-of-the-art research focused on development of rechargeable and reversible energy storage devices such as lithium-ion batteries (LIBs) [1], sodium-ion batteries (SIBs) [2], post-lithium batteries [3] and supercapacitors [4]. To get optimal results, the choice of electrode material is crucial. Carbonaceous materials play an important role as the anode materials for the aforementioned energy storage.

For example, graphite is an excellent material for Li storage, and hard carbons or porous carbon facilitates good Na storage.

Porous solids are characterized by the presence of cavities, channels, gaps, or interstices within their structure, collectively known as pores. These materials exhibit a range of unique properties such as shorter ion diffusion paths, high structural stability, better electronic conductivity, etc. due to their large surface area, adjustable pore size distribution, and high porosity. They can interact with atoms, ions, and molecules not only on the surface but also throughout the bulk of the material. The ratio of total pore volume to the apparent volume of the material is referred to as porosity, a key factor in determining the nature of porous materials. Additionally, the characteristics of porous solids are influenced by the arrangement of pores, as well as their diameter and shape [5]. According to International Union of Pure and Applied Chemistry (IUPAC), porous materials are classified into three categories, such as microporous (pore diameter $<$ 2 nm), mesoporous (pore diameter is 2–50 nm), and macroporous (pore diameter $>$ 50 nm) (**Figure 1**). Hierarchically constructed porous carbon (HPC) materials contain a multiscale porous hierarchy, with porosity and structure extending across various length scales, from micropores to mesopores and macropores. HPCs have immense opportunities in various fields including catalysis [6], separation [7], energy conversion and storage [8], sensing [9], and biomedicine [10]. These materials are of great interest due to their excellent electrical conductivity, high surface area, unique physicochemical properties, multiple levels of tunable porosity, high degree of interconnected porosity, and superior chemical stability [11].

3D carbon architecture has demonstrated significant advantages as electrode materials for lithium-ion batteries (LIBs). As a conductive scaffold, 3D HPCs effectively anchor high-capacity active cathode materials, facilitating improved performance. Their high surface area, abundant porous structure, and continuous conductive networks allow for increased mass loading of cathode materials, leading to enhanced energy storage capabilities. These structural features promote efficient charge transport and ion diffusion, making HPCs highly promising anode material for next-generation LIBs [12]. It facilitates the intercalation of Li^+ ions into the small-scale carbon interlayers, enhancing charge storage. Additionally, HPCs can act as a stable host for Li ions, promoting highly stable cycling and effectively suppressing dendritic Li growth. This property is crucial for improving the safety and longevity of batteries, as dendrite formation can lead to short circuits and battery failure. The use of 3D carbon materials thus enhances both performance and durability in LIBs [13].

HPCs are targeted for various energy applications due to their unique structural properties, which combine high surface area, tunable pore sizes, and excellent

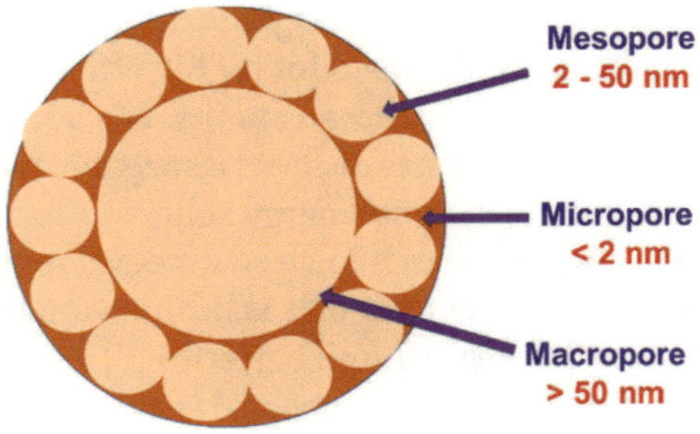

Figure 1.
Represents the type of pore present in a hierarchical hollow porous material.

DOI: http://dx.doi.org/10.5772/intechopen.1008324

electrical conductivity. Superactions, HPC enhances energy storage performance by facilitating efficient ion transport through mesopores while providing high electric double-layer capacitance from micropores, enabling rapid charge and discharge cycles. HCPs are gaining huge attention as the anode material in lithium-ion batteries, sodium-ion batteries and post-Li-ion batteries. Its porous structure offers ample active sites for metal-ion (Li, Na, Al, etc.) intercalation, improving cycling stability and capacity. Furthermore, HPC serves as a catalyst support in fuel cells, where its large surface area and porosity enable effective dispersion of catalyst particles, leading to enhanced catalytic performance [14]. Additionally, its tunable pore sizes make it suitable for gas adsorption and separation technologies, such as CO_2 capture and hydrogen storage, by allowing selective adsorption based on gas properties. Overall, the combination of these advantageous features makes HPC a versatile and effective material for advancing energy storage, conversion, and environmental technologies.

This chapter presents a detailed exploration of the design and synthesis of HPC materials using various strategies. It highlights recent advancements in the use of HPCs and their composites across diverse electrochemical energy storage systems. The chapter also discusses future challenges and research directions in the field of HPC materials.

2. Synthesis of HPCs

HPCs feature a complex architecture due to the combination of various organizational, structural, and geometric pore parameters across different length scales. Key aspects include the number of scale levels ranging from two-stage to multiple stages such as micro-, meso-, or micromacro distributions. Pores can be in various shapes, ordered or disordered, and may have different channel orientations. Understanding these parameters is essential for optimizing the design of these materials for energy storage applications. The concept of "Materials-Properties-By Design" underscores the importance of tailoring materials to achieve desired properties based on their structural characteristics [15]. Block copolymers are crucial precursors for synthesizing HPCs. These copolymers, consisting of two or more covalently linked homopolymers, are categorized into three types based on the number of homopolymers: diblock, triblock, and multiblock copolymers. They self-assemble to form thermodynamically stable nanostructures [16]. One common block copolymer template used for synthesizing HPCs is Pluronic F127. It is a hydrophilic, nonionic block copolymer with a central hydrophobic polypropylene oxide unit flanked by two hydrophilic polyethylene oxide units. Common carbon precursors include phenol, resorcinol, and phloroglucinol, which have one, two, and three hydroxyl groups (-OH), respectively. During self-assembly, these -OH groups interact through hydrogen bonding with the polyethylene oxide units of surfactants [17]. The final step in fabricating HPCs is carbonization, where carbon precursors undergo heat treatment in an inert atmosphere (such as N_2 or Ar). During pyrolysis, the surfactant block copolymers used as structure-directing agents decompose at temperatures above 350°C, leading to pore formation. The carbonization temperature is also important when carbonization is conducted at elevated temperatures (greater than 400°C), and porous carbons often experience significant shrinkage in pore diameter. This can lead to the formation of smaller pores or even the collapse of the porous framework. The morphology of the resulting HPCs is largely influenced by the ratio of precursor to surfactant. Various synthetic strategies are available for synthesizing HPCs, including traditional hard-templating, soft-templating, and self-templating methods.

The widely utilized fabrication method for the synthesis of HPCs is hard-templating or nanocasting strategy. In this method, hierarchical nanoparticles are employed as templates, impregnated with suitable carbon precursor, and carbonization followed by template removal resulting in the formation of ordered mesoporous carbons, more specifically hollow porous carbon, core-shell carbon structure [18]. Examples of common hard templates include SiO_2, ZnO, TiO_2, zeolites, metal-organic frameworks, and nanoparticles. Pal et al. developed hierarchical hollow porous carbon by using nanocasting strategy with silica nanoparticles and a phloroglucinol/1,4-phenylene-diamine/formaldehyde polymer [19]. Although the hard-template approach can efficiently produce ordered, well-defined, and controllable porous carbon frameworks, it is often time-consuming, costly, and labor-intensive (**Figure 2**).

The soft-templating strategy (**Figure 3**) provides an effective alternative for producing hierarchical porous carbon frameworks. In this approach, micelles formed during the reaction process act as templates to generate pores within the carbon structure. During micelle formation, strong interactions such as hydrogen bonding, hydrophobic and hydrophilic interactions, or electrostatic forces occur between the functional groups of the precursors and the templates. These interactions stabilize the micelles, which are then combined with the precursors.

One of the key advantages of the soft-templating method is that all reactant molecules are present in the same medium throughout the process. This uniform environment helps to ensure that the pore structure remains stable and minimizes collapse during critical stages such as gelation, drying, and carbonization. As a result, the soft-templating method can produce well-defined hierarchical porous carbon materials with controlled porosity and structure [20]. Balach et al. achieved the fabrication of micro- and mesoporous HPC monoliths by using copolymer F127 as a soft template and phenol-formaldehyde resin as the carbon precursor [7].

A dual templating approach, combining both hard and soft templates, offers a powerful method for creating HPCs with novel nanostructures. This technique

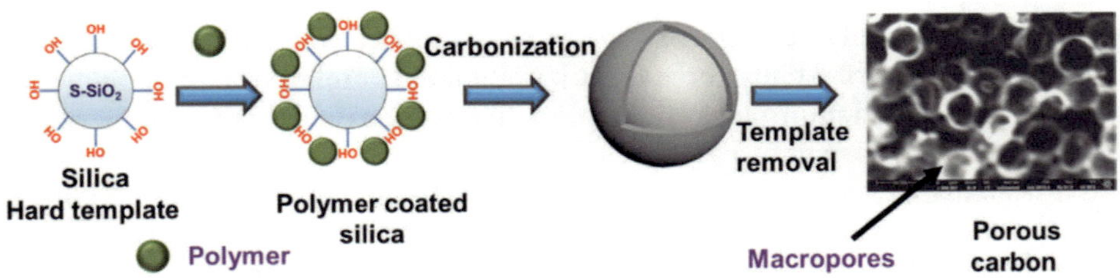

Figure 2.
Schematic diagram of hard-templated synthesis of a hierarchical porous carbon.

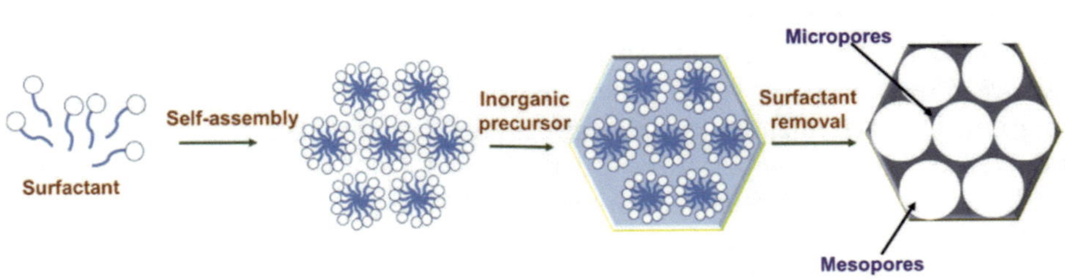

Figure 3.
Schematic diagram of soft-templated synthesis of a hierarchical porous carbon.

leverages the strengths of both hard- and soft-templating strategies to achieve intricate and well-defined porous architectures. In this approach, researchers employ a flow-enabled self-assembly method that integrates hierarchically assembled amphiphilic diblock copolymer micelles with inorganic nanoparticles. The process begins with the formation of micelles from the diblock copolymer, which has both hydrophilic and hydrophobic segments. These micelles can self-assemble into ordered structures due to their amphiphilic nature, creating a soft template. Simultaneously, inorganic nanoparticles are introduced into the system. These nanoparticles interact with the micelles, leading to a coassembly of the micelles and nanoparticles into a hierarchical structure. This self-assembly process is facilitated by the flow conditions, which promote uniform distribution and organization of the micelles and nanoparticles. The result is a hierarchical structure where parallel threads or networks are formed at the nanometer scale. These threads consist of both micelles and nanoparticles, which are organized into a well-defined pattern. When subjected to subsequent carbonization, the micelles and nanoparticles are removed or transformed, leaving behind a porous carbon framework with a novel nanostructure. This dual templating approach allows for precise control over the pore structure and morphology of the resulting HPCs. By combining the self-assembly capabilities of block copolymer micelles with the structural features provided by inorganic nanoparticles, researchers can fabricate innovative HPCs with unique and functional nanostructures. This method opens up new possibilities for designing advanced materials with tailored properties for various applications [21].

Another approach to synthesizing HPCs is the template-free method (**Figure 4**), which includes three main types: self-formation, sol-gel control, and selective leaching. In this method, HPCs are fabricated using directly carbonized, self-generated porogens such as ethylenediaminetetraacetate (EDTA) salts, glycolates, metal-organic frameworks (MOFs), derivatives of MOFs, and organic salts derived from biomass. These materials act as porogens during the carbonization process, creating the desired porous structure in the final carbon product [22]. These methods enable the precise arrangement of organic species at the molecular level, which is essential for regulating the physical and chemical properties of materials. Here, pore-forming agents or porogens are water and alcohol molecules that are released during hydrolysis. Feng et al. developed a direct pyrolysis and template-free method for synthesizing two-dimensional (2-D) Fe/N codoped carbon networks, using 2-D graphitic carbon nitride ($g-C_3N_4$) intermediates obtained from melamine [23].

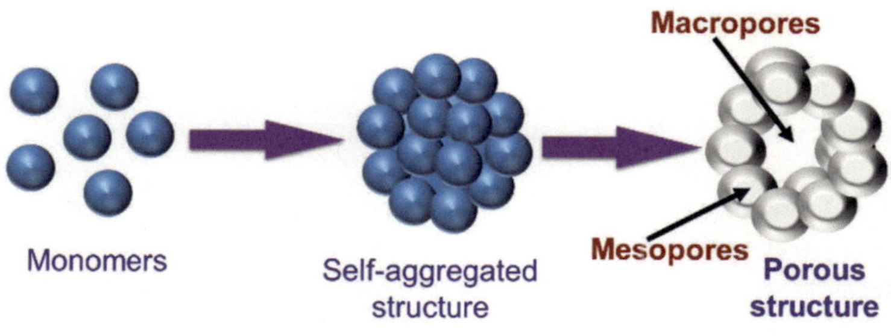

Figure 4.
Schematic diagram of template-free synthesis of a hierarchical porous carbon.

3. Techniques for surface area and porosity analysis for HPCs

Physical gas adsorption/desorption techniques: This analysis is a widely used technique in material science to characterize the surface area, pore volume, and pore size distribution of porous materials like HPCs. This method typically involves exposing the sample to a gas (usually nitrogen or argon) under controlled conditions and measuring the amount adsorbed at various pressures. The sample is then cooled to a low temperature (usually liquid nitrogen temperatures, around −196°C for nitrogen adsorption) and exposed to a known amount of gas. The gas molecules are adsorbed onto the pore surfaces of the material, and the amount of gas adsorbed is measured as a function of the relative pressure (P/P_0). After the maximum adsorption, the gas pressure is gradually decreased, and the desorption isotherm (the amount of gas released) is recorded. The adsorption and desorption isotherms provide valuable information on the pore structure and connectivity. The Brunauer-Emmett-Teller (BET) theory is based on the concept of multilayer adsorption, where gas molecules are adsorbed onto a surface in multiple layers. The BET equation is used to calculate the specific surface area by analyzing the adsorption data within a specific relative pressure range (typically between 0.05 and 0.3 P/P_0). Barrett-Joyner-Halenda (BJH) and nonlocal density functional theory (NLDFT) methods are powerful and sophisticated approaches for analyzing pore size distributions, especially in HPCs. The total pore volume is calculated from the amount of gas adsorbed at a relative pressure near saturation (usually around 0.99 P/P_0), providing an overall measure of the pore volume within the material.

Microscopy techniques: Microscopy techniques provide high-resolution information about the surfaces of materials, allowing for the distinction of surface details at the nanometer scale. Both scanning electron microscopy (SEM) and transmission electron microscopy (TEM) utilize electron beams to interact with the sample, enabling detailed observation of its surface. Additionally, these techniques facilitate compositional analysis of the surfaces, enabling the identification of elemental constituents and the examination of material properties. Focused ion beam scanning electron microscopy (FIB-SEM) combines focused ion beam (FIB) milling with SEM imaging. This technique is particularly valuable for three-dimensional analysis, as it can slice through materials to reveal their layered architecture. Moreover, confocal microscopy, profilometry, and atomic force microscopy (AFM) are among the most valuable techniques for porosity analysis. They provide critical information about the topography and surface roughness of materials, enabling the generation of detailed profiles at nanometric scales. These methods are essential for understanding the intricate surface characteristics of porous materials.

Together, these techniques provide a comprehensive approach to characterizing the surface area and porosity of hierarchical porous carbons, facilitating advances in materials science and applications such as energy storage and catalysis.

4. Importance of the HPCs as an energy storage material

Hierarchical porous carbons are characterized by their well-defined pore dimensions and structures, which significantly enhance their performance compared to conventional porous materials. The distinctive hierarchical structure of HPCs, which includes a combination of macropores, mesopores, and micropores, plays a crucial role in reducing diffusive resistance and optimizing energy storage capabilities [24]. The unique structural design of HPCs optimizes energy storage by utilizing different pore sizes for specific functions. The larger macropores act as "secondary reservoirs" that temporarily store ions. This feature helps to maintain a consistent supply of ions, which is particularly useful during high-demand discharge periods. Mesopores create low-resistance transport channels for ions to travel more easily, facilitating faster movement of ions within the material. This helps reduce the resistance to ion flow, a crucial factor for achieving high power density even at rapid charging/discharging rates. Micropores contribute to the electric double-layer capacitance, an energy storage mechanism based on electrostatic charge separation. This layer is crucial for efficient charge storage, enhancing the material's overall energy density [25].

4.1 Enhanced diffusion and transport

The presence of macropores in HPCs facilitates efficient ion transport by providing larger pathways that reduce diffusive resistance. This is complemented by mesopores and micropores, which offer a large surface area for the dispersion of active sites. The interconnected porosity across these various scales creates a network that shortens the ion diffusion pathway and enhances electron transport. This combination enables rapid and efficient reversible de-intercalation of ions, which is crucial for improving the rate performance of energy storage devices [12].

4.2 Improved energy density and rate capacity

HPC frameworks, enriched with interconnected porosity, enhance both energy density and rate capacity. The large surface area provided by micro- and mesopores allows for increased active site dispersion, contributing to higher energy density. The macropores reduce resistance and enable faster ion movement, which improves the rate capacity. Together, these features enable HPCs to deliver superior performance in energy storage applications.

4.3 Structural stability and cycle life

The low density of HPCs, combined with their ability to accommodate significant volume changes during rapid charge-discharge cycles, enhances their structural stability. This property is essential for maintaining electrode integrity and prolonging the cycle life of energy storage devices. The ability of HPCs to absorb and manage volume fluctuations during cycling helps in sustaining the performance and reliability of the electrodes over time.

4.4 Heat management

The large porous space and interconnections within HPCs also play a role in managing heat during charge-discharge cycles. Effective heat absorption and dispersion are critical for preventing overheating and ensuring stable operation of energy storage devices. By facilitating efficient thermal management, HPCs contribute to the overall durability and performance of these devices.

4.5 Optimized electrode capacity

The combination of low density and high porosity in HPCs ensures that electrodes can achieve fully optimized capacity. The well-designed pore structure not only supports high energy density but also enhances the material's ability to handle rapid charge-discharge cycles, leading to better overall performance [13].

Performance metrics for HPCs are often benchmarked against existing porous carbons to assess their effectiveness and to illustrate the advantages or limitations of the hierarchical design. HPCs typically exhibit increased surface area and pore volume compared to conventional porous carbons, enhancing the number of active sites available for adsorption and charge storage. This optimized pore size distribution allows for efficient ion transport, leading to superior performance in energy storage applications such as supercapacitors and batteries, where HPCs demonstrate higher specific capacitance and energy density [25]. Additionally, HPCs often show improved cycling stability due to their unique structural features, which can better accommodate mechanical stress. However, the complex synthesis required for HPCs can pose challenges in terms of scalability and production costs. Moreover, there may be trade-offs, such as lower conductivity compared to more graphitic carbon materials, which can impact rate capability. Thus, while HPCs provide significant performance advantages, these limitations must be addressed for broader application.

5. Current state of the art

Rechargeable ion batteries have become a critical focus in energy storage technologies due to their ability to efficiently convert chemical energy into electrical energy. Over the past decades, these batteries, including lithium-ion and sodium-ion variants, have attracted significant attention for their high energy density, long cycle life, and broad applications in industries like consumer electronics, electric vehicles, and renewable energy storage. Graphite remains the most widely used anode material for commercial LIBs due to its low cost and excellent cycling stability. Wu's group developed macroporous carbon microballs with mesopores and micropores integrated into the carbon framework, referred to as 3MCM, for use as anodes in LIBs [26]. The 3MCM was synthesized using SiO_2 opal microballs, accumulated from monodispersed SiO_2 spheres as a hard template, with phenolic PF resin as the carbon source. The resulting 3MCM featured an interconnected spherical macropore network (pore size of 250 nm) and a high specific surface area of 966.5 m^2/g, of which 727.5 m^2/g came from mesopores. When examined as an anode material for LIBs, the 3MCM established outstanding rate capability, achieving 248 mAh/g at 20 C, and a high capacity of 1580 mAh/g after 60 cycles at 0.2 C, almost four times higher than the theoretical capacity of graphite anode. The material's performance is attributed not only to its mesoporous structure but also to its enlarged interlayer distance (0.41 nm), which enhances Li-ion diffusion and storage. While the porous structure and enlarged interlayer spacing of carbon materials enhance their lithium storage performance, their intercalation mechanism limits the achievement of high capacity and rate capability. To address this issue, incorporating active materials with alloying or conversion mechanisms, which offer higher theoretical capacities, is a common approach. Alloying anodes like silicon (Si), germanium (Ge), and tin (Sn) are particularly promising due to their high capacities, environmental friendliness, and abundance [25]. Wu and colleagues incorporated germanium (Ge) into a three-dimensional ordered porous nitrogen-doped carbon framework (3DOP Ge@N-C) to develop a binder-free anode for LIBs [27]. Due to the high cost and scarcity of lithium, sodium-ion batteries (SIBs) are emerging as a promising alternative. In the SIBs, graphite's narrow interplanar spacing is unable to effectively accommodate Na-ions, resulting in low specific capacity. Three-dimensional oxygenated hierarchical porous carbon has emerged as a highly promising electrode material, offering unparalleled advantages that make it a strong candidate to replace graphite in alkali-metal ion batteries. Yang et al. created a well-ordered hierarchical porous phosphorus-doped carbon anode using polystyrene spheres as a template, sucrose as a carbon source, and phosphoric acid for doping. This anode featured a pore size of 120 nm and delivered a reversible capacity of 140 mAh/g at 10 A/g [28]. Yang et al. introduced a hierarchical porous C/SiOx anode made from lignite via a one-step carbonization and activation method. The amorphous carbon matrix buffered volume expansion and enhanced conductivity, resulting in a specific capacity retention of 370 mAh/g at 0.1 A/g after 100 cycles and 208 mAh/g after 1200 cycles at 1.0 A/g. Ex-situ analyses revealed the formation of reversible products and irreversible K_2O, demonstrating its potential for high-performance potassium-ion batteries [29]. Lacey et al. first introduced 3D printing technology for developing complex hierarchical porous structures for $Li\text{-}O_2$ battery performance. An additive-free, aqueous carbon-based ink was developed by oxidizing holey graphene (hG) to produce holey graphene oxide (hGO), a nanoporous and hydrophilic material. The 3D printed hGO mesh features multiple levels of porosity, facilitating electrolyte and oxygen gas pathways and enhancing $Li\text{-}O_2$ battery performance. This approach highlights the potential of 3D printing in advancing energy storage devices through hierarchical porous electrode designs [30].

Another important electrochemical energy storage system is supercapacitors. Supercapacitors are known for their ultra-high power density, fast charge-discharge rates, and exceptional cycle stability, retaining their specific capacitance even after nearly 10^5 cycles of operation. Supercapacitors possess high power density, whereas batteries offer high energy density [31]. The charge storage mechanism of supercapacitors follows the formation of Helmholtz double layer at the electrode-electrolyte interface or faradaic charge transfer [32].

Nitrogen-doped hierarchical porous carbon materials are effective as electrode materials for supercapacitors also due to their ability to interact with acidic electrolytes, enhanced by nitrogen doping, which introduces basicity into the carbon framework. The morphology of the carbon and presence of nitrogen atom, well-defined porous structures, further promote both double-layer capacitance and pseudocapacitance. Wu et al. developed a simple method to synthesize high nitrogen-content porous carbon nanorods using cellulose nanocrystals (CNCs) as both a carbon source and template. CNCs were coated with melamine-formaldehyde resin and carbonized at different temperatures under nitrogen to produce N-doped core-shell nanorods (N-MFCNCs). These materials feature a trimodal porous structure (micro-, meso-, and macropores) and a high BET surface area (564.2 m^2/g). The material carbonized at 900°C exhibited the highest capacitance of 352 F/g at a current density of 5 A/g in 1 M H_2SO_4 electrolyte [33]. Pal et al. developed a series of hollow hierarchical carbon structures (HPCS) using spherical silica nanoparticles ($S\text{–}SiO_2$) as templates, aimed at enhancing high-performance energy storage. The synthesis involved coating $S\text{–}SiO_2$ nanoparticles with layers of phloroglucinol/1,4-phenylenediamine/formaldehyde (PPF) polymer, followed by pyrolysis under nitrogen and subsequent removal of the silica core. This process produced hollow spherical carbon structures, HPCS-I, HPCS-II, and HPCS-III, depending on the number of PPF layers. These materials featured uniform spherical morphologies with hollow cores, created by the removal of the SiO_2 template. Among them, HPCS-II exhibited the highest specific capacitance of 592 F/g at 1 A/g, and it retained 444 F/g even at a high current density of 100 A/g, demonstrating excellent charge storage capacity and rate performance, making it highly suitable for energy storage applications [19]. Another interesting strategy is to enhance specific capacitance of doping of HPCs with heteroatoms like boron (B), nitrogen (N), oxygen (O), and phosphorus (P), which has been shown to significantly enhance their electrochemical properties. The introduction of these impurity atoms replaces carbon atoms in the carbon matrix, creating defects that alter the electronic structure of the material. These dopants provide either a single electron or a pair of electrons, facilitating electron donor/acceptor interactions. Moreover, they introduce active redox sites, which play a crucial role in faradaic charge transfer processes, thus increasing the contribution from pseudocapacitance. This modification enhances the overall energy storage capacity, making doped carbon materials highly effective in applications like supercapacitors and batteries [34].

HPCs offer a range of benefits for energy storage devices, including improved ion diffusion, enhanced energy density, and structural stability. Their unique pore structures contribute to better rate performance, longer cycle life, and effective heat management, making them a promising material for advanced energy storage applications.

6. Challenges

1. Scaling up the synthesis of hierarchically porous carbons from laboratory to industrial scale while maintaining quality and consistency is a significant challenge.

2. The use of expensive precursors and templates can drive up the cost of HPCs, making them less competitive compared to other materials.

3. Despite their advantages, HPCs may still need optimization to achieve higher energy and power densities required for advanced energy storage applications.

4. Precisely controlling the size, distribution, and connectivity of pores in HPCs is complex but crucial for optimizing their performance.

5. Ensuring uniformity in hierarchical structures can be challenging, particularly when scaling up production.

6. Integrating HPCs into new and emerging energy storage technologies, such as solid-state batteries or flexible electronics, presents technical challenges.

7. Future opportunity

Streamlining and automating production processes, such as employing 3D printing for fabricating HPCs, can significantly enhance both efficiency and scalability. 3D printing offers several advantages for HPC production.

Tailoring the structure and composition of HPCs can significantly boost their electrochemical performance. By optimizing pore size, distribution, and connectivity, HPCs can achieve higher energy and power densities through improved charge storage and transport. Adjusting material composition and surface functionalization enhances conductivity and interaction with electrolytes, further enhancing performance. Developing composites with other materials, like conductive polymers or metal nanoparticles, can provide synergistic benefits. Fine-tuning synthesis parameters and exploring novel architectures also offer pathways to optimizing HPC properties for advanced energy storage applications.

Leveraging computational models and simulations can significantly aid in predicting and optimizing the performance of HPCs by providing insights into their structural and electrochemical properties. These models help in designing HPCs with tailored characteristics by simulating different configurations and conditions. To ensure the accuracy of these predictions, conducting detailed experimental studies is essential. These experiments validate theoretical predictions and provide empirical data to refine and enhance computational models, bridging the gap between theory and practical applications. Combining both approaches fosters a comprehensive understanding and optimization of HPCs for advanced energy storage technologies.

The exploration and development of HPC materials have significantly advanced the field of energy storage, offering transformative solutions to the limitations of traditional graphite anodes. The introduction of three-dimensional, oxygenated porous structures and innovative doping methods has demonstrated substantial improvements in capacity, rate performance, and cycle stability across various battery systems, including lithium-ion, sodium-ion, and potassium-ion batteries and supercapacitors. As research continues to push the boundaries of material science and energy storage technology, the use of hierarchically porous carbon materials stands out as a revolutionary approach. These innovations not only address existing challenges but also open new avenues for developing more efficient, cost-effective, and high-capacity energy storage solutions for the future.

8. Conclusion

This chapter provides a comprehensive overview of the design and synthesis of HPC materials through various approaches, including hard-templating, soft-templating, and template-free methods. Here, we have discussed these synthesis processes of HPCs through the aforementioned approaches. It reviews the latest advancements in the application of HPCs and their composites as the electrode materials in a wide range of electrochemical energy storage systems, including supercapacitors, lithium-ion batteries, sodium-ion batteries, post-lithium-ion batteries, and hybrid energy storage devices.

Additionally, this chapter highlights future challenges and research opportunities in the field of HPC materials. Key areas of focus include improving the scalability and cost-effectiveness of HPC production, enhancing the performance and durability of HPC-based electrodes, and exploring novel applications and material combinations to advance energy storage technology.

Conflict of interest

The author declares there is no conflict of interest.

Author details

Ananya Pal and Atanu Roy*
The Hebrew University of Jerusalem, Jerusalem, Israel

*Address all correspondence to: atanuroy.physics@gmail.com

References

[1] Degen F, Winter M, Bendig D, Tübke J. Energy consumption of current and future production of lithium-ion and post lithium-ion battery cells. Nature Energy. 2023;**8**(11):1284-1295

[2] Liu T, Yang Y, Cao S, Xiang R, Zhang L, Yu J. Pore perforation of graphene coupled with in situ growth of Co_3Se_4 for high-performance Na-ion battery. Advanced Materials. 2023;**35**(13):2207752

[3] Sun J, Ma L, Sun H, Xu Y, Li J, Mai W, et al. Design of free-standing porous carbon fibers anode with high-efficiency potassium-ion storage. Chemical Engineering Journal. 2023;**455**:140902

[4] Roy A, Ray A, Saha S, Ghosh M, Das T, Nandi M, et al. Influence of electrochemical active surface area on the oxygen evolution reaction and energy storage performance of MnO_2-multiwalled carbon nanotube composite. International Journal of Energy Research. 2021;**45**(11):16908-16921

[5] McCutcheon JR, Ostwal M, Zhang M. Moving beyond passive separations. Nature Materials. 2022;**21**:387-388

[6] Gao S, Jia X, Yang J, Wei X. Hierarchically micro/nanostructured porous metallic copper: Convenient growth and superhydrophilic and catalytic performance. Journal of Materials Chemistry. 2012;**22**(40):21733-21739

[7] Balach J, Tamborini L, Sapag K, Acevedo DF, Barbero CA. Facile preparation of hierarchical porous carbons with tailored pore size obtained using a cationic polyelectrolyte as a soft template. Colloids and Surfaces A: Physicochemical and Engineering Aspects. 2012;**415**:343-348

[8] Kim S, Lee J. Spinodal decomposition: A new approach to hierarchically porous inorganic materials for energy storage. National Science Review. 2020;7:1635-1637

[9] Lin Z, Guo F, Wang C, Wang X, Wang K, Qu Y. Preparation and sensing properties of hierarchical 3D assembled porous ZnO from zinc hydroxide carbonate. RSC Advances. 2014;**4**(10):5122-5129

[10] Meng Z, Wei F, Wang R, Xia M, Chen Z, Wang H, et al. NIR-laser-switched in vivo smart nanocapsules for synergic photothermal and chemotherapy of tumors. Advanced Materials. 2016;**28**(2):245-253

[11] Sun MH, Huang SZ, Chen LH, Li Y, Yang XY, Yuan ZY, et al. Applications of hierarchically structured porous materials from energy storage and conversion, catalysis, photocatalysis, adsorption, separation, and sensing to biomedicine. Chemical Society Reviews. 2016;**45**(12):3479-3563

[12] Pal A, Pal M, Ghosh S, Nandi M. Nanoemulsion technique for the syntheses of N-doped porous carbon nanospheres and their application in energy storage devices. Energy and Fuels. 2022;**36**(19):12285-12298

[13] Wu Z, Hu W, Huang T, Lan P, Tian K, Xie F, et al. Hierarchically porous carbons with controlled structures for efficient microwave absorption. Journal of Materials Chemistry C. 2018;**6**(32):8839-8845

[14] Dessie Y, Tilahun E, Wondimu TH. Functionalized carbon electrocatalysts

in energy conversion and storage applications: A review. Heliyon. 2024;**10**:e39395

[15] Zhang W, Tian Y, He H, Xu L, Li W, Zhao D. Recent advances in the synthesis of hierarchically mesoporous TiO_2 materials for energy and environmental applications. National Science Review. 2020;7:1702-1725

[16] Liu T, Liu G. Block copolymer-based porous carbons for supercapacitors. Journal of Materials Chemistry A. 2019;7:23476-23488

[17] Liang C, Dai S. Synthesis of mesoporous carbon materials via enhanced hydrogen-bonding interaction. Journal of the American Chemical Society. 2006;**128**(16):5316-5317

[18] Fang B, Kim JH, Kim MS, Yu JS. Hierarchical nanostructured carbons with meso-macroporosity: Design, characterization, and applications. Accounts of Chemical Research. 2013;**46**(7):1397-1406

[19] Pal A, Pal P, Nandi M. Carbon structures with hollow internal cavity as charge storage materials to achieve high energy density. Energy and Fuels. 2023;**37**(13):9568-9581

[20] Pal A, Ghosh S, Singha D, Nandi M. Morphology controlled synthesis of heteroatom-doped spherical porous carbon particles retaining high specific capacitance at high current density. ACS Applied Energy Materials. 2021;**4**(10):10810-10825

[21] Han S, Hyeon T. Simple silica-particle template synthesis of mesoporous carbons. Chemical Communications. 1999. pp. 1955-1956

[22] Pal A, Pal P, Das T, Maruyama J, Nandi M. Evolution of an electrochemically inactive metal-organic framework to reticulated porous carbon particles with high supercapacitance. Energy and Fuels. 2023;**37**(13):9582-9597

[23] Feng J, Dou M, Zhang Z, Wang F. Template-free synthesis of two-dimensional Fe/N codoped carbon networks as efficient oxygen reduction reaction electrocatalysts. ACS Applied Materials & Interfaces. 2018;**10**(43):37079-37086

[24] Dutta S, Bhaumik A, Wu KCW. Hierarchically porous carbon derived from polymers and biomass: Effect of interconnected pores on energy applications. Energy and Environmental Science. 2014;7:3574-3592

[25] Fu RW, Li ZH, Liang YR, Li F, Xu F, Wu DC. Hierarchical porous carbons: Design, preparation, and performance in energy storage. Xinxing Tan Cailiao/New Carbon Materials. 2011;**26**(3):171-179

[26] Lian HY, Dutta S, Tominaka S, Lee YA, Huang SY, Sakamoto Y, et al. Curved fragmented graphenic hierarchical architectures for extraordinary charging capacities. Small. 2018;**14**(27):1702054

[27] Wang Y, Luo S, Chen M, Wu L. Uniformly confined germanium quantum dots in 3D ordered porous carbon framework for high-performance Li-ion battery. Advanced Functional Materials. 2020;**30**(16):2000373

[28] Qiao Y, Han R, Pang Y, Lu Z, Zhao J, Cheng X, et al. 3D well-ordered porous phosphorus doped carbon as an anode for sodium storage: Structure design, experimental and computational insights. Journal of Materials Chemistry A. 2019;**7**(18):11400-11407

[29] Yang Z, Zhao S, Jiao R, Hao G, Liu Y, He W, et al. Lignite-based hierarchical

porous C/SiOx composites as high-performance anode for potassium-ion batteries. Energy and Environmental Materials. 2024;**7**(4):e12674

[30] Lacey SD, Kirsch DJ, Li Y, Morgenstern JT, Zarket BC, Yao Y, et al. Extrusion-based 3D printing of hierarchically porous advanced battery electrodes. Advanced Materials. 2018;**30**(12):1705651

[31] Pal A, Das T, Ghosh S, Nandi M. Supercapacitor behaviour of manganese dioxide decorated mesoporous silica synthesized by a rapid sol-gel inverse micelle method. Dalton Transactions. 2020;**49**(36):12716-12730

[32] Roy A, Schoetz T, Gordon LW, Yen HJ, Hao Q, Mandler D. Formation of a CoMn-layered double hydroxide/graphite supercapacitor by a single electrochemical step. ChemSusChem. 2022;**15**(21):e202201418

[33] Wu X, Shi Z, Tjandra R, Cousins AJ, Sy S, Yu A, et al. Nitrogen-enriched porous carbon nanorods templated by cellulose nanocrystals as high performance supercapacitor electrodes. Journal of Materials Chemistry A. 2015;**3**(47):23768-23777

[34] Pal M, Pal A, Pal P, Nandi M. B/P-codoped porous carbon electrode for supercapacitors with ultrahigh energy density. ACS Applied Engineering Materials. 2023;**1**(11):2965-2983